CALIFORNIA NATURAL HISTORY GUIDES

FIELD GUIDE TO ANIMAL TRACKS AND SCAT OF CALIFORNIA

California Natural History Guides

Phyllis M. Faber and Bruce M. Pavlik, General Editors

Field Guide to
ANIMAL TRACKS
and SCAT of
CALIFORNIA

Mark Elbroch
Michael Kresky
Jonah Evans

Illustrated by Michael Kresky and Mark Elbroch

UNIVERSITY OF CALIFORNIA PRESS
Berkeley   Los Angeles   London

This books is dedicated to all the animals, whose tracks we read.

———

University of California Press, one of the most distinguished university presses in the United States, enriches lives around the world by advancing scholarship in the humanities, social sciences, and natural sciences. Its activities are supported by the UC Press Foundation and by philanthropic contributions from individuals and institutions. For more information, visit www.ucpress.edu.

California Natural History Guide Series No. 104

University of California Press
Berkeley and Los Angeles, California

University of California Press, Ltd.
London, England

Library of Congress Cataloging-in-Publication Data

Elbroch, Mark.
    Field guide to animal tracks and scat of California / Mark Elbroch, Michael Kresky,
Jonah Evans ; Illustrated by Michael Kresky and Mark Elbroch.
        p. cm. — (California natural history guide series)
    Includes bibliographical references and index.
    ISBN 978-0-520-25378-0 (cloth : alk. paper) — ISBN 978-0-520-27109-8 (pbk. : alk. paper)
    1. Animal tracks—California—Identification.   2. Tracking and trailing—California.
I. Kresky, Michael.   II. Evans, Jonah.   III. Title.

QL768.E43 2012
591.47′9—dc23

                                                                2011014428

Manufactured in Singapore
19   18   17   16   15   14   13   12
10   9   8   7   6   5   4   3   2   1

The paper used in this publication meets the minimum requirements of ANSI/NISO Z39.48-
1992 (R 1997) (*Permanence of Paper*).{infcir}

Cover photograph: Adult male Cougar tracks in snow in the Mojave National Preserve, southern California. Photo by Mark Elbroch.

The publisher gratefully acknowledges the generous contributions
to this book provided by

the Gordon and Betty Moore Fund
in Environmental Studies

and

the General Endowment Fund
of the University of California Press Foundation.

CONTENTS

ACKNOWLEDGMENTS

We would like to thank the production team at UC Press for their enthusiasm, patience, and attention to detail. Thank you to Anne Hawkins for continued support, patience, humor, and deftly negotiating another contract. We'd also like to thank Liz Snair and Alex Huryn, who shared images of mammal tracks and scats, as well as Rick Truex, wildlife biologist for Sequoia National Forest, William J. Zielinski and Ric Schlexer at the Redwood Sciences Laboratory of the USFS Pacific Southwest Research Station, Preston Taylor, Casey McFarland, and Neal Wight for their support and providing us with access to their track plate collections.

While juggling way too much at UC Davis, many have encouraged me and, at times, even carried me: Heiko Wittmer, Jacob Katz and Kendra Johnson, Bjorn Erickson and Lisa Matsubara, Allison Olliver, Jen Hunter, Maria Santos, Katie Fleer, Phil Sandstrom, Marit Wilkerson, and all those others faithful to Sophia's. Huge thanks to those who support my tracking endeavors: Paul Rezendes, Paulette Roy, Louis Liebenberg, Adriaan Louw, Casey McFarland, George Leoniak, Brian McConnell, Nancy Birtwell, Fred Vanderbeck, and Keith Badger. And last, but not least, thank you to all my family for enduring support and patience.

Mark Elbroch

I'd like to thank Meghan Walla-Murphy, Robbie Olhiser, Stephen Mark Paulsen, Thea Rae Altman, Gary Sieser, Elizabeth Cohen, Preston Taylor, Stephanie Pappas of the Chelonian Science Foundation, the Dautch family, Briggitte Cahill, Arianna Palmisano, Jonathan Weber, and Seston Graham for their support, humor, and generosity in making this book happen.

Michael Kresky

I'd like to thank my wife, Ciel, and our new baby girl, Stella, for supporting my passion for wildlife and tracking.

Jonah Evans

INTRODUCTION
Mark Elbroch

It's not for everyone, this world of little scats and tiny tracks. But for those who are bold, I'd say get some gloves and a jacket and a hat and go out and explore California.

GARY SNYDER IN "THE POROUS WORLD," *A PLACE IN SPACE*, 1995

I arrived at Nettle Springs in the afternoon, a perennial water flow many miles up a dry desert canyon, surrounded by steep, unforgiving slopes of pinyon pine, oak, and juniper. I parked in some shade, filled a water bottle, and strolled up the dry, sandy wash above the springs, where tracks were easy to see. I was looking for an animal to follow. A few deer had crossed the wash early in the morning, but I was hoping for a soft-footed animal, which provided a greater challenge.

Suddenly fresh Cougar tracks were beneath me. The tracks belonged to a mature female cat, and she crossed the wash and traveled northeast into the manzanita, pines, and juniper. After a deep breath to focus and quiet my mind, I began to pursue. The afternoon was fast slipping away, so time was short, and I followed as quickly as my skills allowed. Yet within an hour of starting on the trail, I found myself confused. I looked down at the fresh tracks of a male Cougar—had she really been a he? No, for just around the corner I found her trail again; there were two cats, one following the other. But I was wrong again, for I then noticed the trails of at least three different Cougars converging among the dense manzanita bushes and winding up and over the folds of the desert canyon. I paused to reassess, kneeling down and feeling the large pad of a Cougar's track—that which shows more clearly in difficult tracking terrain.

The mess of tracks could have been a family group. Or perhaps a kill was nearby. That's what I hoped for—a fresh kill—something I could photograph. At the time I knew it didn't really make sense—a mature male and female at the site of a kill, unless one were stealing from the other or they were closely related—but the sun was too low in the sky to give anything much thought, and the trailing demanded all of my attention. I began to jog along the Cougars' trails, circling in on myself and jumping from trail to trail, all the while peering into every shadowy bush and dense growth for the dark telltale mound of a Cougar's cache. Several times I was fooled by the large nests of woodrats. As fast as I tracked, the sun moved quicker, and soon it was turning orange above the mountains to the west.

I followed numerous trails, looped in on myself countless times, and had turned up nothing. I'd not even revealed where any of the cougars had departed the confusion of converging trails. But with sunlight at a premium, I decided to backtrack the female up the wash to see if a piece of the larger story lay behind her. It did. A male lion had been following her down the wash and had cut into the bush farther up the drainage, where I hadn't initially seen his tracks.

I continued to follow her back trail. As the sun moved below the mountain range to the west and the entire canyon was bathed in rosy shadows, I jogged on, reluctant to leave such a beautiful trail. Higher and higher I climbed out of the canyon, but the light was fading fast. Eventually the light was so low, I moved at a crawl and walked in a bent crouch so that my eyes were closer to the ground. I kept losing the trail and having to circle back to find where she had turned. Finally I stood on a steep slope and appraised the areas that I had covered in the valley below and to the north where the cats had spiraled in on themselves. The temperatures were dropping as I began my trip back to camp.

The intimidating stare of a female cougar in a protective stance.

By the time I reached the wash, it was fully dark and stars were twinkling overhead. The simple notes of poorwills sounded in the distance, but otherwise the night was eerily quiet, and my footsteps in the sandy, gravelly wash seemed abrasive and loud. Niggling nervousness began to work its way into my mind, but I threw it off as the usual fears associated with being alone in the wilderness surrounded by fresh signs of Cougars.

Crunch, crunch, crunch, my footsteps echoed along the wash and out into the scrub. I moved quickly to fight off the dropping temperatures and to stop from shivering. The nervousness in the back of my mind hadn't resolved itself completely, and furrowed my brow. Then fear seized me like a slap in the face, and my gut twisted and froze. The moon had just crested the canyon's ridges, and its ghostly light filtered through the bushes and trees, creating shadows and shapes that suddenly appeared menacing and dark. My mind began to imagine a Cougar crouched in every shadow, and I picked up a few good rocks and began to massage my throwing arm to prepare for action in the cold.

"Stop it!" I told myself. "Get control of yourself. It's just the willies. Nothing to worry about. Another mile and a half and I'll be safely back at camp." And so I told myself, "You have permission to be afraid if you see her tracks atop your own." Not likely, I thought, convinced I'd beaten my own mind in the game.

I walked perhaps 50 yards farther before the moon rose high enough to bathe the floor of the wash. Then I saw them. I knelt to study my own tracks made just hours before. And there was no denying it: there were the fresh tracks of a female Cougar atop my own, and she was tracking me. I stood quickly and looked behind me. I was spooked and nervous. I worked my throwing shoulder and rolled my first rock in my hand.

I began to walk quickly down the drainage to the safety of my truck. I stopped with regularity to listen for footsteps behind me, and to look for a Cougar's form in pursuit. I avoided any area where she could attack from above, winding my way down the shadowy wash that seemed to stretch on for eternity. But I arrived safely back at camp, where I decided to sleep in the bed of my truck. With great relief, I slipped into my sleeping bag, thrilled to have shared an evening with another predator.

I awoke some time later with the certainty that I was being watched. The moon was straight overhead and bright. I sat up and peered into the contrasting landscape of dark shadows and reflected moonlight. Nothing but trees, shrubs, shadows, and crisp, cool air. The canyon was utterly silent. I coaxed myself back to sleep, but in what seemed an instant I was wide awake yet again. I gazed out over the edge of my truck bed and beyond the open tailgate at the pines and canyon slopes. Still nothing. Eventually I fell asleep again, and slept through to the first hints of light the next morning.

It was cold, but I was eager to head up drainage to discover where the cat had been when we encountered each other in the wash the night before. So I donned several layers and grabbed a water bottle to fill at the spring. But I stopped at the perimeter of my camp, perhaps 10 meters from where I'd slept. There were her tracks, and she was accompanied by two large kittens, perhaps 10 to 12 months of age. I followed her as she circled my camp, where at intervals she approached where I slept to have another look. She appeared curious. The kittens too, but they never approached as closely as she did. Then she led her kittens down to Nettle Springs for a drink. I continued to follow her as she circled uphill behind the springs and back to where I had slept. She had peered down at me from a height, and then she moved with her family up the hill to the plateau above.

It took the remainder of the day to piece the entire story together, after following her tracks in every direction to relay this story: She'd followed my tracks for half a mile up the drainage to where we'd met in the dark. No doubt she'd heard me coming. She moved off to the north side of the drainage, lay sphinxlike in the shadows of a manzanita bush, and watched me as I passed. From there, she worked her way up the hillside, cutting east as she climbed, paralleling the wash below. Soon she began a more vertical ascent to the north, before dropping into a tiny hidden canyon. It was there she'd left her kittens.

Together the three Cougars dropped back into the wash and followed my tracks toward my camp for well over a mile. She came into my camp, circling, investigating, and then eventually took her kittens to drink. After circling above my camp for another look, they climbed to the mesa above and headed

The author's footprint and the front and hind tracks of the female cougar that followed him back to his camp in southern California. Her front track is below, and the hind track above.

toward Pine Mountain, several miles to the east. It was the direction from which she'd come when followed by the male, so perhaps somewhere near the mountain she'd cached another deer to feed her growing kittens.

What Is Wildlife Tracking?

Wildlife tracking involves the learned skills of interpreting animal tracks and signs, as well as the skills that allow people to follow the subtle signs of beasts over varied terrain. Tracking is about finding animals. It also enriches our outdoor experiences and allows us a real way to engage with wildlife. I never glimpsed a Cougar during my time at Nettle Springs, and I might have overlooked our exchange were it not for the tracks that betrayed the story.

Tracking is many things to many people. This book is about identifying and interpreting some of the physical signs that animals leave in their wake: their tracks, scats, and other signs of scent marking. For those folks also interested in following tracks, refer to *Practical Tracking: A Guide to Following Footprints and Finding Animals* (Liebenberg, Louw, and Elbroch 2010) for an introduction to that subject. Enjoy.

Crisp trail and characteristic drag marks of a mourning dove.

Tracking in Wildlife Conservation, Research, and Monitoring

Wildlife tracking skills are real and can be learned by anyone with patience and persistence. Tracking, however, is a complex skill that requires intelligence and substantial practice to master (Liebenberg 1990; Stander et al. 1997; De Angelo, Paviolo, and Di Bitetti 2010). Readers should also realize that tracking skills are but tools to be wielded for some purpose and not an end in and of themselves; hunting, education, poaching detection, wildlife monitoring, research, and conservation are but some of its varied applications.

Tracking skills are field skills essential to field science, and field science is the foundation upon which our understanding of wildlife is built. Thus, wildlife tracking is fundamental to both field science and wildlife conservation. Furthermore, conservation is our weapon against the insatiable appetites of commercialism and destructive expansion. Conservation constitutes a means of protecting the diversity upon which the health of ecosystems is so dependent. Realize that wildlife tracking has and could continue to contribute much to current and future conservation efforts—such as identifying wildlife corridors and monitoring endangered species. The larger the number of people with wildlife tracking skills, the greater the potential that a strong understanding of natural history and community ecology will guide current conservation efforts. In our minds, this is intelligent conservation.

CyberTracker Evaluator Adriaan Louw discussing the nuances of antelope tracks with Senior Tracker Johnson Mhlanga as part of efforts to sift out those most knowledgeable for employment.

Science and conservation that relies upon wildlife tracking skills to inform their findings and management decisions are only as good as the observers employed to do the work (Evans et al. 2009, De Angelo et al. 2010, Liebenberg et al. 2010, Elbroch et al. [in press]). Numerous studies have and continue to employ observers without the prerequisite tracking skills to record accurate data (Anderson 2001; Karanth et al. 2003; Evans et al. 2009). Conservation efforts built upon poor data are often counterproductive and misleading (Anderson 2001; Galloway et al. 2006; Nerbonne and Nelson 2008; De Angelo et al. 2010). We hope this book will contribute to the training of California's conservationists.

The paramount skill of the tracker, and indeed any field scientist, is humility. Know your limits. You will always find signs that you will be unable to interpret. Become comfortable with saying "I don't know." Regardless of your increasing experience, you will continue to make mistakes. Take comfort in the fact that the top trackers in the world make mistakes with regularity (Stander et al. 1997; Liebenberg et al. 2010); the best trackers in the world are also the first to admit that they made a mistake and to correct themselves.

Also recognize the limits inherent in the interpretation of tracks and signs. Tracking is the interpretation of indirect signs, meaning that we are attempting to recreate what an animal did in the field without ever having seen it. Scientists who employ trackers must be aware of these inherent limitations, and those trackers who work for others must communicate

A beautiful trail of an unknown snake species slithering across exposed dunes.

their limitations coherently. It is detective work, and sometimes we are wrong. That's correct—wrong—and it's no big deal unless we cling to our first interpretation instead of being willing to adapt as new information presents itself. We can only make the best guess with the evidence at hand, and if new evidence points in a new direction, we should throw out our old ideas and make new, better informed ones. Learn not to be attached to your conclusions.

That said, the possibilities in utilizing wildlife tracking skills in the field, and as part of conservation efforts, are limited only by our imagination. Because so few have employed wildlife trackers—or any but the most rudimentary wildlife tracking skills in research and conservation efforts—the potential work ahead of us is both tremendous and exciting (Elbroch et al. [in press]). May tracking and trackers be more greatly appreciated because of this work.

The tracks tell the story. Look closely for the impression of the Bluegill (fish) plucked from the ground by a scavenging Turkey Vulture.

California tarantula and trail. Drawn by Mike Kresky.

Tracking as Storytelling

*To interpret tracks and signs trackers must project themselves
into the position of the animal in order to create a hypothetical
explanation of what the animal was doing. Tracking is not strictly
empirical, since it also involves the tracker's imagination.*

LOUIS LIEBENBERG, *THE ART OF TRACKING: THE ORIGIN OF SCIENCE*,
1990, P. V (IN THE INTRODUCTION)

Tracking is both science and storytelling, and the competent tracker is a practitioner of both. The scientific method involves testing and substantiating hypotheses created in the field, meaning finding and reporting real evidence to support one's claims (for example, the shape of toes as evidence of a Fisher and the known sounds of a bird as evidence the fox was moving nearby).

Storytelling is the interpretation of the scientific evidence with your logical mind and imagination; it is the *reporting* phase of science. Storytelling is also a means by which trackers share their findings in the field with a larger community, whether that consists of family, friends, researchers, or conservation managers. Articulating evidence is an essential skill of the tracker.

Thus, science and storytelling work hand in hand. One cannot be an effective tracker without practicing both skills. Storytelling also allows others to "listen" to what we are finding, so that they may provide us essential feedback that will improve and refine our skills. Trackers receive constant feedback on their skills from wildlife—they either find animals or they don't—but people provide us something more. We are social creatures sharing an increasingly crowded world. Conservation requires teamwork; unified is the only way we will conserve diversity on a global scale.

Tracking Is Possible

Tracking skills are real and amazing. Yet they do not come easily, and they require ample time in the field. Consider this example: Stander and colleagues (1997) observed a Zebra dying of anthrax (blood sampling confirmed the cause), and over the next several days a Lion pride, hyenas, and vultures feeding and disassembling the carcass. They brought a team of four Ju/'huan trackers (Bushman in Namibia) to the area three days after the Zebra had died and asked them to reconstruct what had happened. The trackers investigated the area for two hours before coming to an agreement as to what had occurred. Then they correctly pointed out the spoor of the Zebra and said that the animal was sick and had died from that sickness. They correctly identified all the scavengers and the order in

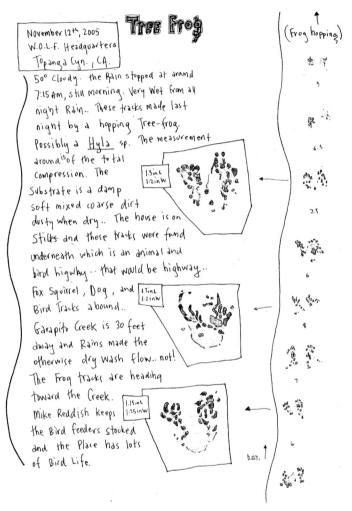

A page from author Mike Kresky's tracking journal.

which they fed. They also correctly estimated the relative chronology of events, including the death of the Zebra.

You, too, can achieve this level of tracking, given persistence and practice. We hope this book provides some of the necessary building blocks and so shortens your journey. Now follow Gary Snyder's (1995) advice found at the very start of this chapter, and "go out and explore California!"

Why Tracks and Scats?

Mark Elbroch

We were faced with a decision at the start of this project. Given the size parameters of this field guide series, we could either present a detailed look at animals' tracks and scats, or use a more superficial approach to a broader diversity of signs—that is, evidence of feeding, beds, and the like. In the end we decided on the former—an in-depth presentation of just several topics. This book is a visual presentation of the tracks, scats, and scent-marking behaviors of California's wildlife.

Interpreting wildlife tracks and signs is challenging detective work; it is an engaging exercise in clear thinking and deductive reasoning. The most reliable tools we have at our disposal to solve the natural mysteries we encounter in the field are animal *tracks* and *scats*. Tracks and scats are the foundation blocks in interpreting all indirect evidence of wildlife, and a detailed knowledge of tracks and scats will aid you in interpreting nearly every natural mystery you encounter.

The presentation of this book also includes numerous citations of work completed by others. They may at first seem obstacles to easy reading, but they serve two powerful purposes: First, they are evidence that tracking has forever been integral to wildlife research and monitoring, and second, they are additional resources for you, the readers, to follow up and seek out should a specific topic grab your interest.

There are without doubt numerous methods to speed you along in your study of wildlife tracking. One of us, Michael Kresky, provides a

Look closely for the California Red-backed Vole remains and the two uric streaks, which are the best evidence that the killer was an American Kestrel.

lively description of one method—track journaling—and through this method he has become an expert interpreter of wildlife tracks and signs. Here he shares tips on this process.

Honing Your Tracking Skills through Track Journaling

Michael Kresky

Tracking Journal

Track journaling builds your knowledge of tracks and sign by making you observe tracks in close detail; your journal also becomes data, meaning scientific records, of what you discover. The physical act of drawing tracks and taking notes on your surroundings sharpens your ability to see clearly and with greater discernment. While you hover over a set of prints, patterns and shapes begin to emerge. We realize that journaling can be intimidating at first, but keep in mind that each journal is highly personal and evolves over time. *Nature Drawing—A Tool for Learning*, by Clare Walker Leslie (1995), can guide you in this process.

Author Mike Kresky journals a set of raccoon tracks under a bridge in Santa Barbara, California.

Tracker Louis Liebenberg taking measurements to create an accurate, life-sized rendition of wolf tracks in central Idaho.

When drawing tracks, do not be daunted by the unknown. Often scrutinizing prints and investigating details with poignant questions will lead to tracks revealing themselves. Even when they do not, it is still valuable to draw the track and revisit your journal at later times.

In the field you need a sketchbook or clipboard, pencils or markers, and a measuring device (ruler or calipers). Begin by finding four to six clear prints that you believe belong to the same animal, and choose a single track to draw. Make sure that the track is between you and the light source. Treat the track and the area around the tracks with deference. This preserves the features of the track as well as the story around the tracks.

An effective track journal includes the date, time, weather, habitat, substrate, and wildlife and human activity in the surrounding area. Draw the track, showing the features of the toes and heel pad or claw marks. With a measuring device, note the track's length and width. The most important thing to remember is to be consistent in your practice. For example, when measuring a track, always include the claws in your measurements. In addition, record whether the track is from a front or hind foot, which side of body, and the direction in which the animal was traveling.

Many folks new to sketching in general are unsure how to use shading to convey light or depth. The choice is ultimately up to you. First, try using shading to relay light—how the sun is falling over and/or across the footprint. Second, use shading to relay depth, meaning that you would shade

The Ghost of Dome Springs

September 8th, 2006... Dry, Hot and yet Summer is losing it's grip... We found this guy's Tracks all the Way up to the Slots at the Top of the canyon...

This scenario was baked dry and hard in a Mud swathe (17.0)

Surrounded by Willow clumps... (16.5) The Animal made a perfect T-step...! (16.5)

* Prey over predator (21)

Left Hind track of a male Mountain Lion...

At the Spring there was a Wall of Cat piss smell!

Puma concolor

Right Fore Track of a ♂ Mountain Lion

"Thick Pugs"

Why'd It make this track?.... what was It doing?.....

3.2 W 2.55 L LSB H

3.45W 3.55L RSB-F

An excerpt from Mike Kresky's journals.

the deepest parts of the footprint the darkest and the shallowest portions the lightest. We would, however, recommend the second method, because shadows created by light sources can distort one's perception of tracks, as we discuss under notes on photography.

Once you have finished drawing the track, step away from it. This gives you a broader perspective, allowing you to see the line of tracks and perhaps discern the gait the animal was using. This perspective offers more information about the interpretation of this animal's behavior. You may notice that the animal was walking, dust bathing, or foraging. After you have drawn an individual track, draw the line of tracks. Measure the distance between the footfalls, again making sure that you are consistent. Although track journaling is a highly focused activity, remind yourself to pay attention to bird song and other animal behaviors. You may be surprised to find that you are near to the animal you are journaling.

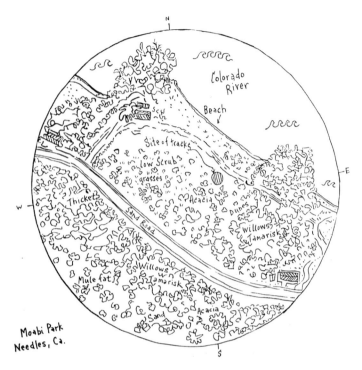

Here Mike Kresky has drawn an area map to accompany his journals of individual tracks and trails.

Maps in Your Track Journaling

It is essential to be familiar with the topography where you track. Topographical maps teach us the salient features of the landscape. Specific maps, such as geological, hydrological, or vegetative, provide different perspectives on the natural history of an area. An awareness of topographical information and patterns in your immediate landscape is invaluable for locating ecotones, corridors, and the animals themselves.

In addition to referencing published maps, create your own maps to hone your tracking skills. Start with a map with a one- or two-mile diameter. Then create a map that covers only several hundred feet. Center this map around your set of tracks. Once the tracks have been drawn, fill in the rest of the map with pertinent features such as vegetation, open space, cover, and topography. Then include bird activity, animal trails and sign, and food sources. This practice reveals consistent trends and patterns on the landscape.

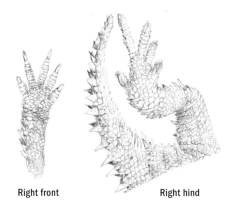

Drawing an animal's feet is also incredibly educational. Here Mike has meticulously drawn the feet of a horned lizard.

Right front   Right hind

Field Guides and Nature Documentaries

A collection of field guides is an indispensible tool for finding the answers to many questions that arise from tracking. A diverse library gives you access to information not readily available in the field and builds your ecological understanding. When available, use numerous resources to identify and research a track, feather, or chewed acorn. Cross-referencing builds a more dynamic picture of the specimen you are studying, lending a more objective view. Sources beyond field guides can add valuable information to a reference library. Scientific journals, natural history magazines, news articles, local papers, and the Internet can all broaden your knowledge of tracking and ecology.

While field guides aid in identification of specific species, they also help to develop search imagery for species you have yet to encounter in the field. By casually flipping through the pages of field guides you develop a familiarity with species yet to be encountered.

Nature documentaries expose you to animal movements and behaviors. To assist in your concentration on animal movement, mute the sound track and if available, use slow motion. These movies allow you to watch animals move over a long period of time, thus enabling you to better interpret track scenarios.

In addition, watching animals in videos will help you see how track patterns are made. Watch how an animal's feet fall as it walks, trots, or gallops. Pay attention to the transitions between gaits. You also learn to interpret sitting, foraging, and other behavioral patterns.

Tracking Exercises

Below are a variety of tracking exercises that will help push your tracking skills beyond what can be learned through books and journals.

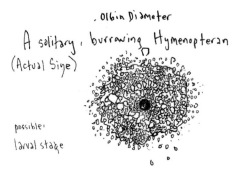

. 016in Diameter

A solitary, burrowing Hymenopteran
(Actual Size)

possible:
larval stage

Mike Kresky's journals are filled with varied and wonderful discoveries, and it is his insatiable curiosity that has made him an excellent tracker.

MAKING SPECIES LISTS A species list compiles all the animals that occur in a given area. For example, if you wonder what mammal tracks you may encounter near Baker, California, obtain a field guide that covers this region, such as *Mammals of California* by Jameson and Peeters (2004). Flip through every page, and determine via the range map if that species lives in or near Baker. If so, write that species on your species list. Now, when you find a track in Baker, you can be relatively certain the animal is on your list. Organize the list by keeping the animals in their taxonomic families. Other useful species lists to make are trees, birds, plants, amphibians, and reptiles. This exercise prepares you for what you may encounter in the field.

THE SEVEN PERSPECTIVES OF A TRACK This exercise is based on seven separate drawings of a single track and sharpens your observation skills by teaching you to see from different perspectives. Find a clear print: (1) Draw only the outline of the track. (2) Draw the geometric shapes—such as triangles or rectangles—that compose the overall track. (3) Sketch the variation in texture within the track. (4) Next, study the different colors in the track. Even if you do not have colored pens or pencils in the field, take the time to notice the colors in and around the track and note them as best you can. (5) Draw all the edges of the track. Edges include all the areas in and around the track where soft or hard transitions exist. (6) Draw the variation in value within the track, which is the difference in lightness and darkness of the track. (7) Combine all the previous drawings into one.

DRAWING A SCENARIO Find a patch of ground where there are tracks of multiple species. Sketch all the tracks, and notice how they intermingle and overlap. Begin to ask yourself which tracks are older than others. By sequencing the events you begin to see the story unfold. Keep in mind how weather, sunlight, and shade affect tracks. Seton (1958) provides excellent examples of drawing track scenarios.

STAKING OUT A TRACK LINE Begin by finding a series of 20 or more tracks made by the same individual. Then, using wooden skewers or any consistent flagging device you have brought into the field, place one skewer per

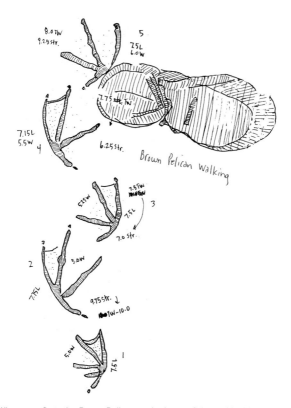

Who came first, the Brown Pelican or the human? Journal by Mike Kresky.

track. Once stakes have been placed, step back from the trail and observe the revealed pattern. In staking out a track line, you suddenly begin to see how the animal moved across a landscape. You see the gait and how the landscape affected the animal's movement. This exercise is excellent for visualizing how animals move.

LEARNING TO MOVE LIKE AN ANIMAL The more time you spend interpreting the tracks of animals, the more you will recognize the importance of gaits. The following activity is done in conjunction with a field guide or nature documentary that shows the gaits of different animals. Begin by choosing one animal to mimic. Study how that animal walks and where each of its feet land in that gait. Practice walking like that animal so that your own hands and feet fall in the same positions of the animal in the field guide. From there you can work on other gaits of that animal. When you are comfortable with one species, move on to others.

The tank tread made by a traveling Desert Tortoise in the dunes of the Mojave National Preserve in southern California.

BUILD A TRACKING BOX A tracking box can be any size, from a small transportable shoebox to an area large enough for a human to run through. Once the box is ready, fill it with "good" substrate. The easiest medium is play sand purchased at a local hardware store, but you can experiment with local soils as well. Once you have chosen your substrate, tamp or trowel the surface until it is smooth and firm.

Tracking boxes afford many activities for the beginner as well as the more seasoned observer. For instance, place your tracking box near bird feeders or where you know animals frequent. By providing a controlled substrate you can track in any season regardless of circumstances, even if you live in a place where seasonal weather or field conditions make tracking difficult. Once an animal has walked through your box, analyze the tracks as you would in the field. Photography, journaling, and track casts can all be used in conjunction with a tracking box.

The tracking box is also an excellent tool for recording the tracks of specific species. For example, capturing small mammals or live insects and running them across the substrate allows for focused analysis of elusive tracks. To further hone your skills, step beyond the sandbox and into the wild. Smooth out substrates along your daily routes, and monitor the movements of animals in your area.

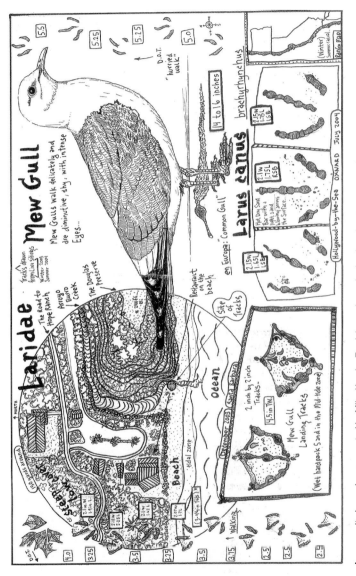

A composite journal page created by Mike Kresky. Track journals are both educational and beautiful.

Tracker's Tools: Noninvasive Monitoring

Jonah Evans, Mark Elbroch, Michael Kresky

One of the most attractive aspects of wildlife tracking is that it is non-invasive and provides a rich, detailed record of wildlife presence, habits, and behaviors without the necessity of seeing or handling them. Wildlife tracking provides us cost-effective research methods, and indirect signs allow both hunters and biologists to invest their time in the appropriate place at the appropriate time if they want to catch their quarry.

Below is a brief introduction to several complementary research methods that you might be interested in experimenting with. We highly recommend that you also read *Noninvasive Survey Methods for Carnivores* by Long et al. (2008), which provides both a more detailed description of all of the following methods and an introduction to the science behind their use in wildlife research and monitoring.

Track Casting

Making a cast is one of the time-honored ways of creating a permanent record of an animal track. Because of the ephemeral nature of tracks in the field, a track cast invites you to revisit tracks over and over again. The three-dimensional aspect of the cast allows for more in-depth analysis of foot morphology. Creating a track cast collection is a tangible way to increase your exposure to animal tracks when not in the field. Track casts are also great for teaching tracking in a classroom or workshop setting.

The simplest and most cost-effective medium to use is plaster of Paris. It pours easily and dries quickly in most weather conditions. Latex, dental mold, or paraffin are used less often but are also effective. The following tips are for making casts with plaster.

The trail and burrowing mound made by a mole crab. Their trails are common at low tide in the intertidal zone.

MIXTURE In a suitable container (either a plastic cup or Ziploc bag), mix a ratio of about two parts plaster to one part water in a quantity to fill and cover your track. The consistency of the mixture is of greater importance than the exact ratio and should be thick like a milkshake. Thinner, more watery mixtures pick up better detail, but they are more fragile and may have a chalky consistency when dried. We would also recommend that you add the water slowly. You can always add more, but if you add too much, you must hope you have additional plaster in your pack to thicken your solution sufficiently to use.

POURING THE PLASTER Many tracks are delicate, and pouring plaster directly onto the track can damage them. It is a good idea to hold your mixing implement about an inch above the track to break the fall of the plaster, which will then gently fill the track. Once the plaster has been poured, it is a good idea to tap or shake the surface with your mixing implement to encourage any air bubbles to surface and break. This also helps move the plaster into the deepest cavities of the footprint, like the fine marks made by claws.

DRYING Well-mixed plaster on a hot day in the sun and on dry substrate can be dry enough for transport in less than 20 minutes. In cold, wet conditions it may take a few hours. The cast should be fully cured in 24 to 48 hours.

OTHER TIPS A barrier of some kind placed around the track to hold in the plaster can improve the appearance and thickness (strength) of the cast. You can easily make such barriers by cutting out strips from plastic containers of various sizes.

As soon as water is added to plaster, a chemical reaction occurs that releases a small amount of heat. This is problematic for snow conditions, because the plaster melts the snow and misshapes the cast. The solution is a product called SnowPrint Wax, which is a wax that can be sprayed into the track before pouring the plaster. This product prevents the snow from melting.

You may also come across a track preserved in silt mud or clay. You can carefully cut and lift out such a track and preserve it in a dry area. Unlike casting, this preservation technique provides a more pristine first generation of the track; even better, it is not subject to the variables of casting.

Whether you are casting or cutting a track out of substrate, be aware of your impact on the land. Also realize that casting is potentially messy and may involve toxic materials. Clear and concise procedures regarding setup and cleanup of casting are imperative.

Photography

Photographs are a quick way to record tracks and are an excellent tool for learning. Inexpensive digital point-and-shoot cameras are capable of taking decent-quality track photographs and can be carried in a shirt pocket. The drawback of a compact camera is the small sensor, which often equates to high noise levels in low-light situations (common when photographing

Cameras are also excellent tools for recording rare animals. Mark Elbroch snapped this photo with a cheap flip phone to record the tracks of a rare Wolverine atop Castle Peak in the northern Sierras.

tracks). High-quality digital single-lens reflex cameras can produce professional-quality images but are bulkier, more expensive, and require more experience and skill to use correctly. Regardless of which camera you choose to carry on your tracking expeditions, your photographs will become a powerful tool for learning and documenting your observations.

Photographs allow you to analyze details that you may have overlooked in the field. By keeping your photos organized and labeled, you can compile a library of different species that you can revisit and compare with each other. You can also enlarge your photos and look for details you may have missed in the field.

Tips for the Field

FILL THE FRAME When taking an image of a single track, position the camera directly above the print and fill as much of the frame as possible. This may require being on your knees and elbows, but photos taken from a standing position rarely have enough detail.

STEADY THE CAMERA Supporting the camera on a tripod is the best method of track and sign photography, because a tripod allows you to use the necessary large F-stops to capture an entire image in focus even in shaded area or low-light conditions. It is possible to take good-quality images by hand when there is adequate light. If you are hand-holding a camera, it may help to stabilize your elbows against the ground or some other object. A general rule of thumb is to use the smallest aperture possible that allows you to use a shutter speed fast enough to avoid blur. Traditionally the slowest hand-holdable shutter speed is the numerical equivalent of the focal length of the lens. Thus, if you are using a 100-mm lens, never use a shutter speed slower than 1/100th of a second.

SHADE THE TRACK Unless the sun is at a very low angle (sunrise or sunset), it is preferable to shade the track. Direct, overhead sunlight removes contrast and may create shadows that distort the shape of the footprint.

Common Poorwill tracks in shade. The penny is ¾ in. (1.9 cm) in diameter and is a useful scale you might have in your pocket.

The deeper, darker impressions in the snow are the trail of a hunting Ermine tracking down a deer mouse that left the smaller, lighter tracks grouped in fours. Photo by Jonah Evans

USE A SCALE Try to place a ruler or other object of known size in the image. In some situations, the absence of a scale can make it much more difficult to identify the track. This is especially important when recording data and documenting the presence of a rare or controversial species.

TAKE SEVERAL IMAGES The great thing about digital cameras is that taking photos is free. Take several photos of each track and vary the shutter speed, aperture, zoom, and other camera settings. When you get home you can select the best image and delete the rest. Taking a few extra photos is well worth the effort when compared to the disappointment of finding that the one photo of a track you took is blurry.

Carnivore Scent Stations

Carnivore scent stations are a fancy name for a human-made track trap. Scent stations were developed by researchers to attempt to maintain a probability of detecting an animal across locations, time, or both. One of the challenges of using tracks in research is that substrates vary, and it is easier to find tracks in some areas than in others. Thus, if few tracks are recorded in an area, this might mean that few animals inhabit the area, or it might be a result of poor tracking conditions.

Scent stations include a scent lure and either an artificial tracking surface such as sand or gypsum, or an enhanced surface obtained by raking an area or sifting local soils. Researchers place scent stations across a landscape and use visitation rates of different species to estimate a variety of parameters including relative and actual abundances. See Long et al. (2008) for a detailed presentation of methods and scientific designs using both scent stations and prepared tracking substrates without baits.

Sooted Track Plates

Sooted track plates are another method developed by researchers to equalize the probability of detecting an animal's tracks. While more difficult to set up than scent stations, they have the advantage of providing a record of the track that can be carried out of the field for further inspection. Track plates come in a variety of forms to suit different species, but the basic principle is the same: an animal is lured across a board or "plate" covered with soot, chalk, or some other powdery substance, and then a sticky paper (contact paper) catches the soot from the bottom of the foot. The resulting two-dimensional tracks can be incredibly detailed or a blur of black dust; they are also very different from the three-dimensional tracks we see in earth and snow. For that reason, we have provided here a brief guide to common tracks you might catch when you do this in California, which begins after page 30. This resource was compiled from three resources: (1) tracks caught as part of a southern Sierra Nevada carnivore monitoring program managed by Rick Truex, wildlife biologist for Sequoia National Forest, (2) the collection of William J. Zielinski and Ric Schlexer at the Redwood Sciences Laboratory of the U.S. Forest Service Pacific Southwest Research Station, and (3) several private libraries, primarily Mark Elbroch's, but with generous contributions by Preston Taylor, Casey McFarland, and Neal Wight.

Track plate methodology has been particularly refined for marten and Fisher detection. Herzog et al. (2009) developed a technique to identify individual Fishers from footprints on track plates and thereby found a noninvasive means of gathering capture-recapture data for quantifying population estimates. Slauson, Truex, and Zielinski (2008) determined that total track length (as defined in the accompanying visual) was an effective method to determine the sex of marten tracks collected on track plates in California. In the Sierra Nevada, the cutoff point is 30.75 mm—larger measurements are males, smaller are females. In the northern Humboldt region, martens are smaller, and the cutoff between the genders was 29.70 mm. Slauson and colleagues (2008) also presented a more complex equation to determine the sex of Fishers from their tracks in California: $(1.046 \times 13 \text{ height}) + (1.011 \times \text{total length}) + (0.361 \times 13–14 \text{ width})$. Refer to the illustration to see how to measure these variables. The cutoff point for this equation is 62.17; larger numbers are males, and smaller are females.

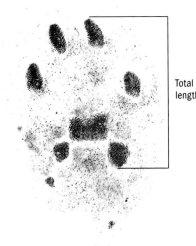

Total length

Measurements used in determining gender of marten, and one of the measurements needed to determine gender of Fisher tracks as described by Slauson et al. (2008). Note track is not to scale.

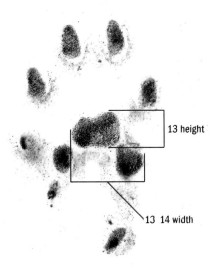

13 height

13 14 width

Additional measurements useful in determining gender in Fisher tracks on soot plates. For this animal, we calculated 60.93, or female. As described in Slauson et al. (2008). Track is to scale.

For additional information on track plates, see Long et al., *Non-invasive Survey Methods for Carnivores* (2008), Zielinski and Kucera, *American Marten, Fisher, Lynx and Wolverine: Survey Methods for their Detection* (1996), and Halfpenny et al., *Track Plates for Mammals—A How-To Manual and Aid to Footprint Identification* (2010).

Camera Traps

There has been an abundance of relatively cheap remote cameras flooding the commercial market in recent years. Hunters often use them to monitor trails, so that they know exactly when and where large bucks prowl. Researchers use them to document the presence of numerous species, especially rare and elusive ones.

Camera traps are a wonderful complement to traditional tracking skills. They provide a visual confirmation to our interpretations, and provide a way to learn more about tracking. If there is a mystery path created by some animal in your backyard, perhaps a well-placed camera trap can teach you what species created it.

DIGITAL VERSUS FILM The primary disadvantages of digital remote cameras remain shorter battery life and shutter delay, but these are being resolved with each new generation of cameras. The benefits of not needing to pay for film developing, as well as the ability to instantly see photos (and make adjustments as needed), make digital cameras a better choice for most situations.

SOOTED TRACK PLATES

PLATE 1 Sooted Tracks: Actual Size

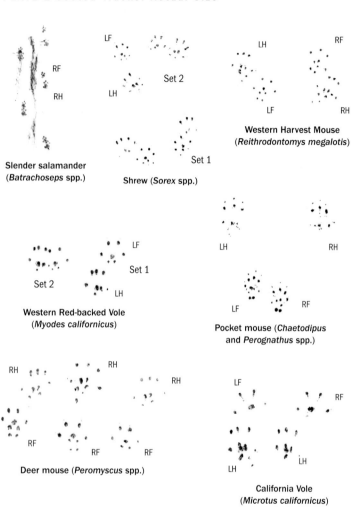

Slender salamander
(*Batrachoseps* spp.)

LF
Set 2
LH
Set 1
Shrew (*Sorex* spp.)

LH
RF
LF
RH
Western Harvest Mouse
(*Reithrodontomys megalotis*)

LF
Set 1
LH
Set 2
Western Red-backed Vole
(*Myodes californicus*)

LH
RH
LF
RF
Pocket mouse (*Chaetodipus*
and *Perognathus* spp.)

RH
RH
RH
RF
RF
RF
Deer mouse (*Peromyscus* spp.)

LF
RF
LH
LH
California Vole
(*Microtus californicus*)

RF
RH
Rough-skinned Newt
(*Taricha granulosa*)

RF
LH
California Newt
(*Taricha torosa*)

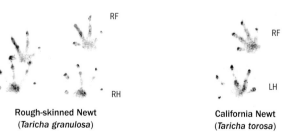

PLATE 2 Sooted Tracks: Actual Size

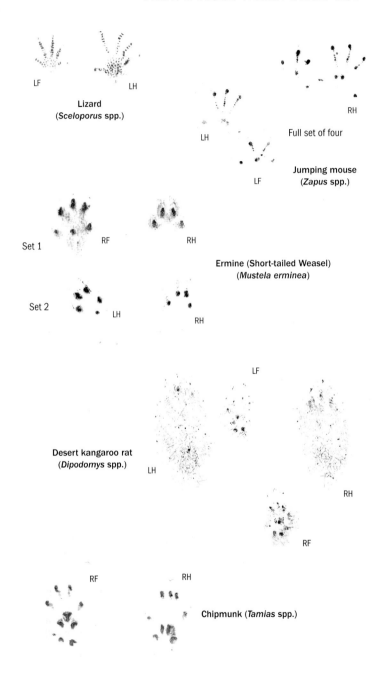

LF

LH

Lizard
(*Sceloporus* spp.)

RH

LH

Full set of four

LH

LF

Jumping mouse
(*Zapus* spp.)

Set 1

RF

RH

Ermine (Short-tailed Weasel)
(*Mustela erminea*)

Set 2

LH

RH

LF

Desert kangaroo rat
(*Dipodomys* spp.)

LH

RH

RF

RF

RH

Chipmunk (*Tamias* spp.)

L = LEFT R = RIGHT F = FRONT H = HIND

PLATE 3 Sooted Tracks: Actual Size

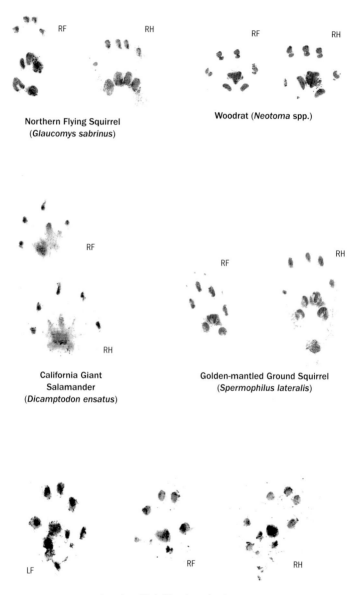

RF   RH

Northern Flying Squirrel
(*Glaucomys sabrinus*)

RF   RH

Woodrat (*Neotoma* spp.)

RF

RH

California Giant
Salamander
(*Dicamptodon ensatus*)

RF   RH

Golden-mantled Ground Squirrel
(*Spermophilus lateralis*)

LF   RF   RH

American Mink (*Neovison vison*)

PLATE 4 Sooted Tracks: Actual Size

California Ground Squirrel
(*Spermophilus beecheyi*)

Western Spotted Skunk
(*Spilogale gracilis*)

Western Gray Squirrel
(*Sciurus griseus*)

L=LEFT R=RIGHT F=FRONT H=HIND

PLATE 5 Sooted Tracks: Actual Size

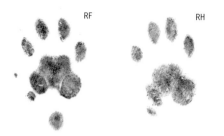

RF   RH

Ringtail (*Bassariscus astutus*)

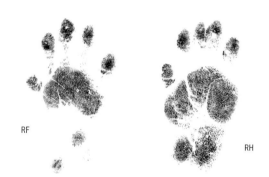

RF   RH

Striped Skunk (*Mephitis mephitis*)

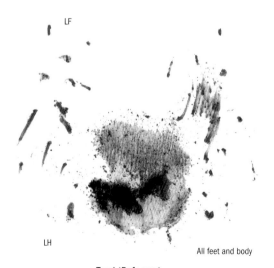

LF

LH

All feet and body

Toad (*Bufo* spp.)

PLATE 6 Sooted Tracks: Actual Size

Snowshoe Hare
(*Lepus americanus*)

RH

RF

RH

L=LEFT R=RIGHT F=FRONT H=HIND

PLATE 7 Sooted Tracks: Actual Size

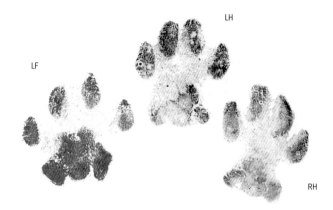

LH

LF

RH

Domestic Cat (*Felis catus*)

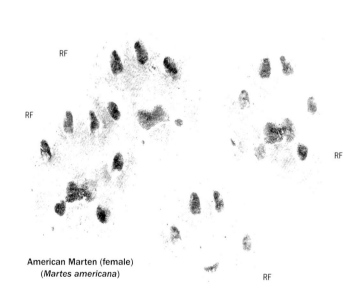

RF

RF

RF

RF

American Marten (female)
(*Martes americana*)

PLATE 8 Sooted Tracks: Actual Size

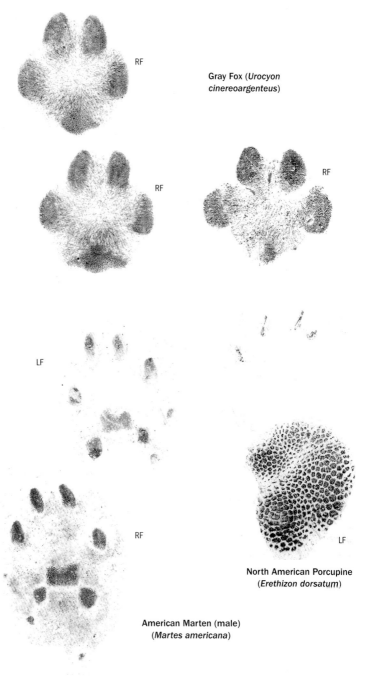

Gray Fox (*Urocyon cinereoargenteus*)

North American Porcupine (*Erethizon dorsatum*)

American Marten (male) (*Martes americana*)

L=LEFT R=RIGHT F=FRONT H=HIND

PLATE 9 Sooted Tracks: Actual Size

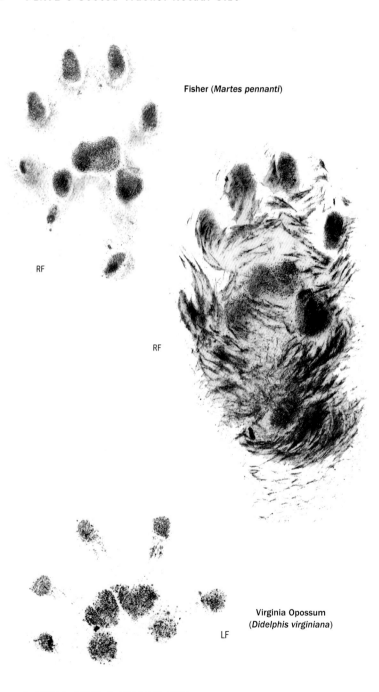

Fisher (*Martes pennanti*)

RF

RF

Virginia Opossum
(*Didelphis virginiana*)

LF

L=LEFT R=RIGHT F=FRONT H=HIND

PLATE 10 Sooted Tracks: Actual Size

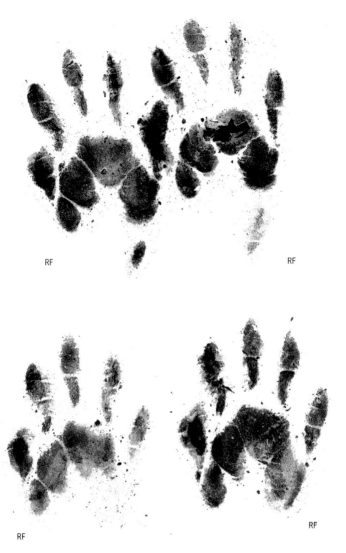

RF

RF

RF

RF

Northern Raccoon (*Procyon lotor*)

PLATE 11 Sooted Tracks 75% of Actual Size

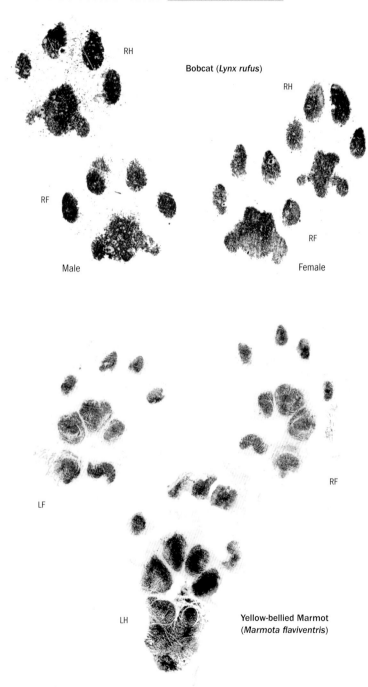

Bobcat (*Lynx rufus*)

RH

RH

RF

RF

Male

Female

LF

RF

LH

Yellow-bellied Marmot
(*Marmota flaviventris*)

PLATE 12 S o o t e d T r a c k s 75% of Actual Size

RH

RF

Cougar (female) (*Puma concolor*)

MAMMAL TRACKS AND
TRACK PATTERNS

Mark Elbroch

To better visualize and understand tracks on the ground, we must study whole animals, starting with their feet. Feet make footprints, and footprints are the building blocks of track patterns. Here we begin with the morphology of mammal feet, then move on to study the morphology of mammal tracks, then look at how mammals move, and finally delve into the science and interpretation of mammalian track patterns on the ground.

Foot Morphology

Bones

Thousands of years ago, the first mammals had five toes on each foot (Hildebrand and Goslow 2001). Over great lengths of time, through evolution and specialization, feet and legs evolved and became more varied. The foot structure of extant shrews is believed to reflect that of the first mammals.

The skeletal structure of the forefeet consists of the carpal bones, metacarpal bones, and phalanges, while that of the hind feet consists of the tarsal bones, metatarsal bones, and phalanges. The original mammals were plantigrade, meaning that they were supported by all the bones of their feet while they were moving, whereas today's mammals have diversified in structure and function. In general, modern plantigrade animals have relatively short limbs and prefer to walk, because the construction of their feet is not well adapted for jumping or running for long distances. In contrast, long-distance runners have long limbs, and the area of their feet that is in contact with the ground is as small as possible. To obtain a firm grip on the ground, the foot must exert the greatest possible pressure to dig into it. Because pressure is equal to force per unit area, the contact area must be as small as possible to ensure the greatest possible pressure. For example, in deer the second and fifth toes are reduced to form dewclaws, and their weight is supported on the third and fourth toes; thus, they have long legs and small feet relative to their size.

Bears and humans are examples of extant plantigrade species. Digitigrade species, such as felids (cat family) and canids (dog family), support themselves on the distal ends, or heads, of the metacarpal bones and the phalanges of the forefeet, and the distal ends of the metatarsal and phalanges of the hind feet. Even-toed unguligrade species such as deer and bison, and odd-toed unguligrade mammals such as horses, walk on the distal phalanges of the third and fourth toes, the equivalent of our toenails.

In animals with five well-developed toes (or digits), they are numbered from 1 to 5 beginning with the innermost toe, that which corresponds with the human thumb (see the accompanying illustrations of woodrat and river otter feet on pages 33 and 34). The third toe in most mammals is the longest and largest, followed in order of size by the fourth, second,

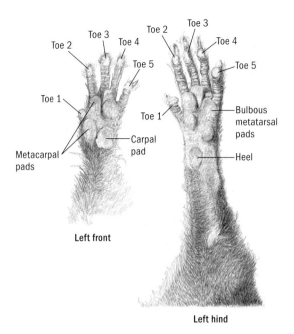

Left front

Left hind

Big-eared woodrat feet.

fifth, and first. In some mammals the first toe is tiny and only makes a weak impression, or none at all. If all five toes are showing and the shortest toe is on the left side of the footprint, the track was made by a right foot. If only four toes are showing and the shortest toe is on the left side of the track, then it was made by a left foot.

With evolution, however, some species developed structures and gaits that required different foot structures. Bang and Dahlstrom (1972) note that many species, such as deer, began to evolve longer bone structures to aid in running, while at the same time losing toe 1 altogether to become more streamlined.

Pads

Tough horny layers of skin cover elastic masses of connective tissue that protect bones and other foot structures from rough ground. In most mammals these pads are naked (rabbits are an exception), and the spaces between these pads are filled with fur (some species have completely naked feet, like Striped Skunks). The thick toe and metacarpal or metatarsal pads that compose the "palms" of each track are covered in sweat glands and deposit scent with each step (see the accompanying illustrations of river otter feet on page 34).

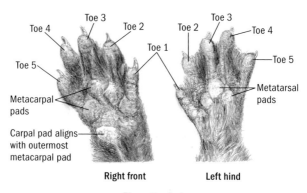

Toe 3 | Toe 2 | Toe 2 | Toe 3
Toe 4 | | | Toe 4
| | Toe 1 |
Toe 5 | | | Toe 5
Metacarpal pads | | | Metatarsal pads
Carpal pad aligns with outermost metacarpal pad

Right front   **Left hind**

River otter feet.

The pads at the tips of the toes are called the digital pads. The pads that form the "palms" of footprints are called the metacarpal pads on the front feet and the metatarsal pads on the hind feet, and they correspond to the respective bones that they protect. In many animals, such as Bobcats and Coyotes, the metacarpal pads of the front feet and metatarsal pads on the hinds are fused to form one large pad. In addition to these, some animals have one or two additional pads to the posterior. The larger pad on the front feet is the carpal pad and covers the carpal bone. When there is a second, it is another metacarpal pad linked with a reduced and sometimes redundant inner digit (see the figure of the woodrat on page 33). In mammals, carpal pads are found only on the forefeet.

Feet

If the front end of an animal is larger and heavier than its hind end, the forefeet will be larger and broader than the hind feet. The forefeet support the head, chest cavity, and forequarters of the body, which are often heavier than the hindquarters. Some mammals, such as bears, rodents, and otters, have larger hind feet and more massive hindquarters than forequarters.

The forefeet are generally rounder in shape than the narrower hind feet, because the forelegs are almost perpendicular to the ground while the hind legs are positioned such that the heel is held at an angle to the ground. A cylinder that is perpendicular to a plane has a circular cross section, while a cylinder that meets a plane at an angle has a larger, elliptical cross section.

Evolutionary adaptations of feet reflect specific types of locomotion, as well as their use in some animals as tools or weapons. Predators have soft pads for stealth, and some have sharp claws to hold down their prey, or short, blunt claws that aid in traction. Some animals have claws that are tools for digging, and others that are adapted for grooming. Feet adapted

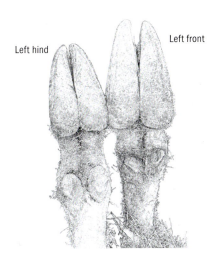

Left hind

Left front

Male Bighorn Sheep, approximately seven years of age. Note the huge disparity in size between the front and hind feet.

for soft, muddy ground require a large contact area for support. Sharp, pointed hooves can dig into soft, sandy terrain to obtain a firm grip, while hooves must be rounder to better grip firm ground. Aquatic mammals often have webbed toes or a stiff fringe of hair to increase the surface area of their feet and therefore the resistance with which they pull themselves through the water. Animals adapted to arboreal environments have sharp claws to dig into the bark of trees, or opposable joints or toes to grasp branches.

The more massive ungulates, such as American Bison, have broad, round hooves, while lighter deer have slender, narrow hooves. Very sharp, pointed hooves are an adaptation for speed, especially in soft, sandy substrate, and act like spikes to prevent slipping. Pronghorns, which prefer open terrain, have to rely on speed to escape being captured, and have sharp, pointed hooves. While habitat specialization results in foot structures with certain advantages, it also places the same animal at a disadvantage in alternative habitats. For example, the large, webbed feet of beavers are ideal tools for aquatic living but cumbersome obstacles while negotiating land at speed.

Track Morphology

Most mammal tracks are composed of claws, digital pads, metacarpal or metatarsal pads (the palms), carpal pads and other heel structures, and the negative space in between these structures (see the figure on page 36). To start, study beautiful footprints in damp, slightly muddy earth, wet sand, a

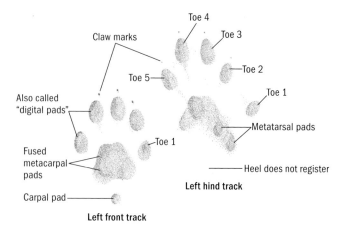

Toe 4

Claw marks

Toe 3

Toe 2

Toe 5

Toe 1

Also called
"digital pads"

Metatarsal pads

Toe 1

Fused
metacarpal
pads

Heel does not register

Left hind track

Carpal pad

Left front track

North American River Otter tracks.

thin layer of loose dust on firm substrate, or a thin layer of fresh snow. In these conditions you will be able to see clear, complete tracks and better understand them. An individual animal's footprint will vary according to its age, mass, sex, condition, and the substrate in which it steps, and these nuances can only be recognized once you have enough experience with a species to know what a "normal" track might look like.

The specific details of track morphology for different species are covered in depth in the species accounts at the end of the book. However, as an introduction to the topic, consider these eight key questions to species identification and some examples of why they might be useful:

1. *How asymmetrical is the track?* Track symmetry is the comparison between the right and left sides of the track. In a perfectly symmetrical track, the left and right sides align and match perfectly. However, nearly every mammal track is asymmetrical to some degree, and it is the degree of asymmetry that will be useful.

   For example, the front tracks of Bobcats are often much more asymmetrical than the hind tracks. The opposite is true for river otters, where the front track often appears more symmetrical than the hind track.

2. *How many toes can you see in the track, and are they all on the same plane?* Shrew tracks have five toes on both front and rear feet, and they all tend to register in footprints. Hares have five toes on their front feet (although the innermost toe is tiny and easily overlooked in the track or does not register at all) and four toes on their hind feet. Canid and felid tracks tend to register four distinct toes, even though they have five toes on their front feet; digit 1 on each front

The loping tracks of a river otter. From right to left, right front, left front (below)–right hind combo, left hind. Compare the relative sizes and symmetries between front tracks and hind tracks. Study the shapes of the digital pads and claw marks.

foot sits on a higher plane on the inside on the leg and registers when they run.

3. *What is the shape of the digital pads?* Teardrop-shaped toes are characteristic of felids, especially females, as well as the digits in mink and smaller weasel tracks. In Fisher tracks the bulbous toe pads are *separate* from the metacarpal or metatarsal pads, while in raccoon tracks the cigar-shaped toes tend to be connected to the metacarpal or metatarsal pads.

4. *Can you see nails?* Look closely, because sometimes nails register as tiny pricks in the ground, as in Gray Foxes; in other instances they are massive and difficult to miss, as in American Beavers. Are they sharp or blunt? Are they coming straight out from the toes, or do they curve down from a location on top of the toe? In deep snow, Bobcat tracks often show claws, which register in the wall of snow in front of the track, on a plane *higher* than the digits. Domestic Dog tracks tend to have large, blunt nails in comparison with the sharp, thin nails of Coyotes. The outer nails of Coyotes also often register so close to the inner toes that they are overlooked.

5. *What is the overall shape of the metacarpal or metatarsal pads (palms)?* In some animals, including canids and felids, the metacarpal pads have fused together to form one palm pad. Characteristics of felid tracks include an anterior (leading) edge with two lobes, termed bilobate, and a posterior edge with three. The palm pads of felid tracks also tend to fill the overall track. This means that, in terms of area, they compose a larger proportion of the track than do the palm pads in tracks of canids. In both canids and felids, the palm pads are much larger on the front feet than the hind.

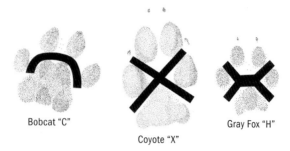

Bobcat "C"

Coyote "X"

Gray Fox "H"

A comparison of the negative space in the front feet of three carnivores.

6. *Can you make out a heel in the track?* All mammals in the order Rodentia, from mice to beavers, excepting the porcupine, register two round pads behind their metacarpal pads in their front tracks. One is a carpal pad on the heel and the other a modified metacarpal pad. In the front tracks of weasels, Ringtails, and bears, a carpal pad can be found, and in some mammals the heels of the hind feet register only as impressions.

7. *Can you see hair in the track, and if so, where?* Rabbits lack naked pads and have completely furred feet. Red Foxes have very small toe pads surrounded by lots of hair, and Striped Skunks have no hair between their toes at all.

8. *What is the shape in between the toes and palm pads, here called the negative space?* Look for an "X," "H," or "C" shape in felid and canid tracks. The front tracks of Gray Foxes and Domestic Dogs tend to show an "H," while those of Red Fox and Coyote show an "X." Look for a "C" in the front tracks of cats.

The Effects of Substrate

Substrate is a catchall word for what an animal has stepped in, whether it be sand, mud, snow, or grass. The depth of substrate, which is reflected in the depth of the print, has an enormous influence on the appearance, size, and shape of the track, as well as on how the animal moves. In shallow substrates, like moist, hard sand, animals move easily and therefore tend toward their normal gaits. In deep or slippery substrates, animals tend to slow down. The conditions of different substrates in which an animal may step are infinite and thus create great challenges for the tracker.

The toes on soft feet are rounded in soft ground but will spread out on firm ground and appear larger and different in shape. On very hard ground only the tips or edges of hooves may show, or only the claws of padded feet. Movement and activities also change the shape of a track. For example, sliding feet may give the impression of elongated toes, and

To illustrate the effects of substrate and behavior on tracks, compare the following three photos. In this picture, a Bobcat has stepped in firm mud and its resulting tracks are clean, tight, and easily recognizable.

Here a Bobcat is walking normally in soft, deep dust that obscures the edges of the tracks, making them more difficult to identify.

These are the tracks of a Bobcat stepping in soft, moist mud, and the cat responded by spreading its toes (resulting in splayed tracks) and in some instances, extending its claws to improve traction. Three radically different sets of tracks, all made by the same species.

twisting and dragging feet may partially obliterate track features. The forefeet and hind feet may also be superimposed, so that the toes of one foot may be confused with those of another. Running animals may also splay their feet. Certain species have tremendous control in how much they spread their feet. For instance, cats walking in soft mud or wet snow splay their feet and leave tracks nearly twice as large as those left when on firm ground.

When loose windblown sand has accumulated in a footprint that was made in damp or wet sand, it is sometimes possible to carefully blow away the loose sand to reveal the features of the footprint underneath. Footprints in mud may in fact be preserved for quite a long time underneath a layer of loose sand or leaf litter. When leaves are covering the track, or even when the animal has stepped on top of leaves, they can be carefully removed to reveal the track. When studying tracks in loose sand or other difficult substrates, you should try to visualize the shape of the footprint before the sand grains, snow, etc., slid together to obliterate the well-defined features.

What Else Can Tracks Tell Us?

While species can be identified by characteristic features, there also exist individual variations within a species. These variations make it possible for an expert tracker to determine the sex as well as estimate an animal's approximate age, size, and mass. A tracker may also be able to identify a specific individual by its footprint. For example, Stander et al. (1997) tested these abilities in four Ju/'huan trackers in Namibia, and they correctly answered 557 of 569 questions. The Ju/'huan team correctly identified the species that made the track in 100 percent of tests. They correctly identified the relative ages of Cheetahs and Leopards (that is, cub, juvenile, young adult, adult) 100 percent of the time in 30 tests, but the relative age of African Lions only 34 out of 39 times (87 percent). Their mistakes were all in calling a subadult animal a full adult. They correctly identified the sex of Lions in 100 percent of 39 tests, Cheetahs in 12 of 13 tests, and Leopards in 16 of 17 tests. It is also in principle possible to identify an individual animal from its track. It may have a unique way of walking or a particular habit that distinguishes it from other individuals. The Ju/'huan trackers correctly identified the individual Lion, Leopard, or Cheetah by its tracks in 30 of 32 tests (96.4 percent).

Determining the Gender of an Animal

Many mammal species show characteristic sexual dimorphism, meaning that one sex is significantly larger and heavier than the other. In the case of California's mammals, it is likely to be the male that is larger. Sexual dimorphism is especially prevalent and blatant in California's carnivore species, including canids, felids, pinnipeds, mustelids, and bears, where adult males are always the larger sex. In species where dimorphism is

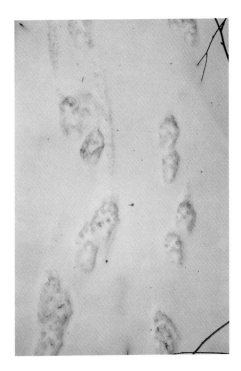

On the left are the tracks of a mature male Fisher, and on the right the much smaller tracks of an adult female.

especially pronounced, such as Cougars and Fishers, the tracks and trails of adult males and females can quite easily be differentiated. In other species, such as river otters and skunks, sexual dimorphism is minimal, and therefore track size alone is not as useful a character in differentiating males from females.

What follows are some useful track characters to consider in determining the sex of select carnivore species. With considerable aid from numerous African trackers who have determined how to sex Leopard and Lion tracks, one of us (Mark Elbroch, unpublished data) has developed this list to aid researchers in identifying the sex of Cougars from their tracks. He has identified the sex and relative age correctly in 100 percent of trials with wild Cougars ($n = 28$), where the animal was subsequently caught as part of larger research efforts to verify the interpretation. The ability to reliably determine the sex of Cougars from their tracks is widespread in Cougar houndsmen that hunt them for sport, with depredation permits, or for a living, and this skill would be a valuable one to document further and quantify.

When studying the following track characters, you should build a case for a particular gender; that is, determine how many characters indicate female and how many male. Sometimes a track yields conflicting evidence, and the gender cannot be determined with certainty.

The left front and hind (lower right) tracks of a young adult male Cougar, approximately three years of age, which means he still had a bit more growing to do.

Track features: (1) Tracks of adult male Cougars are larger than those of adult females, and the areas of the metacarpal and metatarsal pads are also larger. Measure the width of the metatarsal pads (palm) in the hind track, and use a cutoff of 50 mm to determine sex. This is a very reliable tool in determining the sex of Cougars, especially when the other characters described here provide conflicting evidence. This method has also been used by other researchers (Shaw 1979; García et al. 2010). Only the odd female will approach this cutoff, but adult male palm pads typically start at 53 mm wide and can be much larger. García et al. (2010) also report that in captive Cougars they used to study tracks, the "shape of the heel pad in males can be described as relatively narrow in the middle, while that of females is more extended in this zone." (2) The toes of males in both front and hind tracks are blockier than females; the toes of females are more slender and more teardrop-shaped. (3) The front tracks of males are almost always wider than long and are at least as wide as long; the front tracks of females are often narrower than long or as wide as long. (4) There is a smaller area of negative space between the metacarpal pads and the toes in the front tracks of males than those of females. It is as if the palm fills the track more in male tracks than in female ones. (5) The negative space between the metatarsal pads and the toes in the hind tracks of males is much smaller than that found in hind tracks of females; typically male palm pads nearly touch the toes, and in females there is a significant rectangular gap. (6) There is less space between the toes of the hind tracks in males than in females. (7) The hind track of males is generally narrower than long, but not to the degree that the hind tracks of females are.

Front and hind (above) tracks of an adult female Cougar, with penny for scale.

Trail characteristics: (1) The strides of females (measured from left hind track to left hind track, or right to right) average about 36 in. and the strides of males about 40 in. (Boone Smith, personal communication). (2) The width of walking trails is much wider in males than in females.

Elbroch has applied these same characters to other species, including Gray Foxes, Ringtails, Bobcats, American Minks, and Fishers. Preliminary results are very positive yet not so much so that we can say with certainty that they work all of the time. We encourage you to start gathering the data needed to develop these skills further, and the track and trail measurements from known-sex animals, so that we can begin to compile useful parameters for species in North America.

In hooved species, the relative weight difference between an individual animal's front and hind ends is proportional to the discrepancy in size between its front and hind tracks. In Bighorn Sheep and Mule Deer, where males are adorned with large horns or antlers, their front ends are much larger than their hind ends to support this additional weight; their feet and footprints reflect this discrepancy, and their front tracks are much larger than their hind ones. Female Bighorn Sheep and Mule Deer also have larger front tracks than hind tracks, but only slightly larger, and their front and hind tracks are much more similar in size. Females of both species carry their weight more evenly distributed

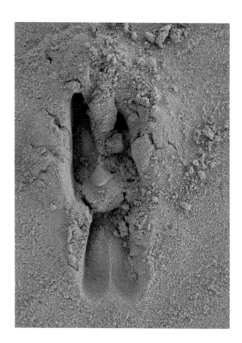

Carefully note the large disparity between the length and width of the front track above and hind track below in a mature Mule Deer buck.

between their front and hind ends than do their male counterparts. It is the larger difference between the length—and more often the width—of front and hind tracks of males that helps trackers determine an animal's sex. That difference in size between front and hind tracks of females is much smaller.

The sexes may also be distinguished by association. The track of an adult in close association with a juvenile is probably that of a female with her young. Nursery herds may be identified by the presence of several young, or the absence of young may indicate a bachelor herd. When a species is gregarious, a solitary individual will probably be an adult male.

Determining the Relative Age of a Mammal

The relative age of an animal may be indicated by the size of the feet. The hooves of young deer will have sharper edges, while older individuals may have blunted hooves with chipped edges. Younger individuals with padded feet may have more rounded pads. Some animals have specific breeding periods. If it is known at what time of the year an animal is born and you know the relative growth rates for the species, a reasonably accurate estimate of a young animal's age can be made from the size of the track.

Here the larger track of the adult female Cougar on the left is easily differentiated from her five- or six-month-old kitten on the right. Through association, determining the sex of the adult is made easy, and with practice (and ample opportunity to see the animals responsible for making the tracks), determining the relative age of the younger animal from the size of its footprint is possible.

Track Patterns

Competent trackers complete two tasks in interpreting a series of animal tracks. First, they analyze the pattern of tracks on the ground so as to visualize how the animal was moving. Second, they interpret the behavior—the meaning in what they are seeing. The first task is more systematic and scientific, while the second is more speculative, relying more upon imagination. Let's take each in turn.

Gaits and Track Patterns

A *gait* describes the way in which an animal is moving; it is *not* a description of a specific track pattern. There will be numerous track patterns for each gait, depending upon the speed and behavior of the animal, as well as the anatomy and morphology of the species.

We must also be aware of the complications introduced by language differences. Trackers around the globe use different words to describe the same thing. Some people prefer words that describe track patterns on the ground, while others prefer terminology that describes the way an animal was moving (the gait). This book presents the vocabulary most widely accepted by trackers across the globe, and that which was used by Eadweard Muybridge (1957) in his pictorial presentation of *Animals in Motion*. This vocabulary provides you with visual information about how an animal is moving—which is crucial in envisioning and becoming that animal in advanced levels of tracking (Liebenberg 1990). But understand that no terminology is better than another. What's important when communicating with others about gaits and trails is that you are all envisioning the same motions.

Here we separate gaits into three categories:

1. *Walks and trots* in which the front feet fall at consistent, rhythmic distances from each other (created by keeping the spine straight and allowing the momentum to be driven by motion in the legs);
2. *Lopes, gallops, hops, and bounds* in which the front feet land in alternating short and long distances from each other (created by stretching and contracting the spine, in addition to moving the legs); and
3. *Bipedal* gaits in which animals move on only their hind legs.

Walks and Trots
Walking
Walking is a slow gait in which each foot moves independently, and at no point during a cycle of footfalls does the animal lose contact with the ground. We will arbitrarily begin with the right hind foot in our example. The right hind leg moves forward, and just before touching down, the

A walking Polar Bear.

right front lifts up and moves forward. For a moment, two feet are off the ground, and then the right hind touches down. The right front continues forward and then touches down. The left hind moves forward, and just before it touches down, the left front picks up and moves forward. For a second time in the cycle of footfalls there are only two feet in contact with the ground, and then the left hind touches down. The left front continues forward and then touches down. Immediately the cycle begins again, and the right hind picks up and moves forward. Rhythmically, this would sound like "1, 2, 3, 4, 1, 2, 3, 4, 1, 2….," where each number is an independent footfall. Note that the rhythm is continuous—that is, without breaks or pauses.

When an animal is walking, its hind foot may land in any relation to the front track made by the front foot on the same side of the body. Remember, the right front foot picks up and moves forward before the right hind foot touches down. For this reason, the hind foot may land exactly where the front foot had been placed—that is, atop the track made by the front foot, called a *direct register*—or even touch down beyond the front track, called an *overstep*.

A good portion of time during each cycle of footfalls, only one leg is lifted from the ground, which allows for three feet to support the animal while in motion. These three legs act similarly to the legs of a tripod, which is a sturdy arrangement that efficiently balances heavy objects, including wide animals with short legs.

Walking is common among almost all the animals. For many wide-bodied animals, such as beavers, porcupines, and bears, it is their most common method of moving. It is also the common gait for deer, elk, antelope, and all members of the cat family. Other species walk when exploring or while traveling in deep substrates, like snow, to save energy.

A variation of the walk is the stalk. In the stalk, only one limb moves at a time, but the order in which the feet move is the same as for the walk. The right hind foot moves forward and touches down. The right front foot moves forward and touches down. Then the same sequence occurs for the left hind foot, followed by the left front. The resulting trail is an *understep* walk, which means the hind tracks in each pair register behind the front tracks.

There are numerous variables to consider when interpreting speed from a series of tracks in a trail, but a general rule holds true for walking gaits. As an animal walks faster, its hind track will move over and beyond

The direct-register walk of a female Cougar moving along a hiking trail in shallow snow.

The overstep walk of the same female Cougar, in which the hind tracks pass over and register beyond each front track.

The peculiar walking gait of the raccoon.

the front track in each pair. Therefore, an understep (where the hind track lies behind the front track) is probably a slower gait than a direct-registering walk where the hind lies on top of the front, and both are probably slower than an overstep walk, where the hind track registers beyond the front track. A fast walk is also called an *amble*. Remember, there are other variables to consider as well, such as depth of substrate, and the anatomy and morphology of the specific animal you are tracking. However, the speed can generally be inferred, up to a point, by considering the placement of the hind track in relation to the front, for an animal can only walk so fast before it must change gaits to increase speed further.

RACCOONS AND 2 × 2 WALKS. Raccoons walk in distinctive fashion. The front and hind legs on each side of the body move nearly simultaneously (called a pace), and they really stretch forward with their front limbs. In the resulting track pattern the tracks are paired, and each pair consists of a front or hind foot from one side of the body and the opposing front front foot from the other side of the body. This means that if in one pair you see a left front and a right hind track, the very next pair of tracks will be of the right front and the right hind foot. Refer to the illustrations.

This variation of the walk can also be altered by adding or subtracting speed. Look at each pair of tracks, and note where in relation to the front track the hind track registers. When raccoons walk slowly the hind track falls behind the front track, and when they walk quickly the hind track falls beyond the front track in each pair. The distinctive fast walk of the raccoon is trotlike in speed and unique to this species.

Trotting

Faster trots are easily differentiated from slower walks, in that a front and hind leg on opposing sides move together, as if joined by a cable. Rather than each foot moving independently, two legs move simultaneously, and there is a moment during each cycle of footfalls when the animal becomes airborne, completely losing contact with the ground. This vertical component is easily seen in any canid, from foxes to Domestic Dogs; there is a little bounce in their common gait. This means that as the right hind foot shifts forward in the air, so does the left front foot. Just before the right hind and left front are placed down, the right front and left hind feet push off, maintaining forward momentum. Then, just before the right front and left hind touch down, the animal pushes off with the right hind and left front feet. The cycle begins all over again. Rhythmically, the beat is a

continuous, unbroken 1, 2, 1, 2, 1 …, where each number is two diagonally placed feet landing simultaneously.

The common gait of canids (members of the dog family) and many voles is the trot. Many other species also use trots as a travel gait, including badgers, Mule Deer, American Black Bears, and members of the cat family. Bighorn Sheep tend to trot on flat ground, where they are at greater risk of predation.

The classic 2 × 2 pattern created by a walking raccoon is a distinctive character that aids in their identification.

A trotting wolf.

A Domestic Dog illustrates the side trot.

Direct-register trots, in which hind tracks are superimposed on front tracks of the same side of the animal, are common in many species. Hind feet land exactly upon the recently made front tracks, as the front foot and opposite hind pick up just before the alternate pair touches down. The forward momentum of the animal carries the hind foot directly over the track just made by the exiting front foot on the same side of the body.

The perfect straight line created by a trotting Coyote—moving away from the photographer.

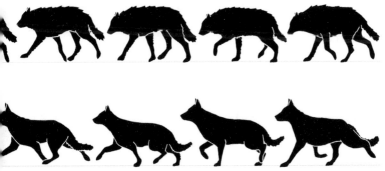

As with walking, speed estimates can be inferred by the position of the hind foot in relation to the front track. Direct-register trots are theoretically slower than overstep trots, in which hind tracks in a given pair register beyond the fronts. Overstep trots are achieved by species in different ways. The hind foot can only move forward so far before it collides with the front foot of the same side. Animals typically overcome this obstacle in two ways. Canids tend to use the *side trot*, or crab, and trot with their entire body at an angle; they kick out their rear end to one side. In this way the hind feet pass to one side of the front feet so as to move at a faster pace.

The side trot of a Red Fox in light snow. On the left side of the trail are all of the hind tracks, and on the right are all of the front tracks.

A Caribou illustrates the straddle trot.

When an animal is using a side trot, all the front tracks appear on one side of the trail, and all the hind tracks on the other side; the hind tracks are also slightly forward of each front track in each pairing. This unique gait and track pattern is easy to find along Coyote, Wolf, and Red Fox trails; Gray Foxes very rarely side trot.

Another option is to kick each hind leg out to either side of the front legs, and this is called a *straddle trot*. All the canines use this gait, but it tends to be for short sections of trail and most often is a transition from a direct-registering trot to a side trot. However, Gray Foxes use this gait very often, as do Mule Deer and several shrew species. The final option to successfully overstep trot involves a longer airborne time, allowing the hind feet to glide over the front tracks with forward momentum, as seen in lizards and occasionally other animals. This last option is rare in mammals but is sometimes incorporated into dominance displays and aggressive interactions.

Some lizards can trot in the technical sense, and achieve high speeds, while other reptiles and amphibians use a gait that simultaneously exhibits characteristics of both trotting and walking. Most turtles and tortoises are so wide and heavy that they move each foot independently, as described under walking above, so as to benefit from tripod support from their remaining three legs. Others, including salamanders, walk with their opposing (diagonal) front and hind limbs in synch—as described under trotting above. However, unlike trotting animals, they never leave the ground when moving in this manner, and their movements often appear exaggerated as they curve their spines side to side to increase speed. Note, this is why it is impossible for a salamander to completely direct-register walk; the right front foot is still on the ground when the right hind comes in behind, and therefore blocks it from moving any farther forward. For this slower gait, we suggest the term *diagonal walk* to differentiate it from trotting. In a diagonal walk the animal remains in contact with the ground yet is propelled forward by diagonal limbs moving simultaneously.

Lopes, Gallops, Hops, and Bounds

Lopes and Gallops

Lopes and gallops are very similar gaits, and the fastest gaits for mammals. During lopes and gallops, each foot lands independently of the others but in rapid succession. During both gaits the animal becomes momentarily

A Caribou in a slow lope.

A loping Wolverine—also called a 3 × 4 lope.

A galloping Bobcat.

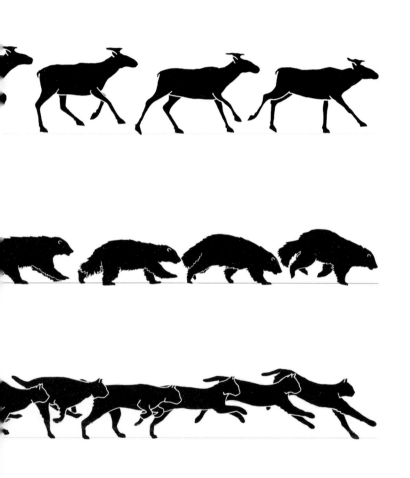

airborne, just after pushing off with the front legs, but during gallops there is a second point at which the animal is in the air, just after pushing off with the hind legs. Not only is it difficult to catch this second, short flight when watching an animal move, but it is also challenging to distinguish between lopes and gallops when interpreting track patterns on the ground.

There are two possible sequences in which the feet touch the ground during lopes and gallops. If they land in a circular fashion—left front, right front, right hind, left hind—it is called a rotary lope or gallop. If the order does not circle the body but instead cuts across the body—left front, right front, left hind, right hind—then it is called a transverse lope or gallop. Rhythmically there is a three- or four-beat sequence for lopes, followed by a pause: 1, 2, 3, pause, 1, 2, 3, in which the number 2 is the overlapping strike of the second front and the first hind foot to touch down, *or* 1, 2, 3, 4, pause, 1, 2, 3, 4, pause, 1, 2 …. Gallops produce a four-beat sequence, followed by a pause: 1, 2, 3, 4, pause, 1, 2, 3, 4, pause, 1, 2 … Neither lopes nor gallops produce continuous rhythms, as are found in walks and trots.

When looking at track patterns on the ground, a lope typically becomes a gallop when both hind feet land beyond both front feet, but this is not always the case. If the order of tracks on the ground in a single set of four prints is front, hind, front, hind, then it is a lope. If the order is front, front, hind, hind, then it is more likely a gallop.

Similar to walks and trots, the placement of the hind tracks in relation to the front tracks betrays speed. As the pair of hind tracks moves beyond the pair of front tracks, this indicates a faster lope or gallop. Also note the distance between the groups of four tracks. In general, the longer the distance spanned by four tracks in a series and the shorter the length in between groups of four tracks (called the stride), the faster the animal is moving. At all-out speeds, some mammals will leave track patterns that look to the casual observer like trots, in that tracks are placed regularly and in a straight line. But each mark is a single track, rather than two.

Three particular lopes are characteristic of the mustelids, or weasel family. They are the 3×, 3 × 4, and 2 × 2 lopes. The 3× and 3 × 4 lopes are rotary lopes, in which a front and hind on the same side of the body may land in the same space, giving the impression of only three tracks in a set (the 3×), or land so that all four tracks are clearly evident (the 3 × 4). Hildebrand and Goslow (2001) show that weasels still have a front foot on the ground when the first hind foot touches down, and therefore this gait is a true lope.

A Fisher in a 2 × 2 lope.

The 2 × 2 lope is a transverse lope, although other than the order of footfalls it is very similar in body mechanics to the 3 × 4 lope. The same fluid arcing motion is used for both gaits. What is unique about the 2 × 2 lope is that the front feet pick up and the hind feet land directly upon the front tracks—creating a trail of paired tracks, where each set of two prints is actually a set of four, the fronts registering first and the hind feet registering directly on top of them. Meadow voles and smaller shrews also use this gait in deep snow. Based upon Hildebrand and Goslow's research on the 3 × 4 lope, we assume that a front foot is still in contact with the ground when the first hind touches down. Should we be wrong, then this gait would technically be a gallop.

Hopping and Bounding

The hopping and bounding gaits of rabbits and many rodents are different from lopes and gallops, in that the hind feet land and push off simultaneously, or nearly so. This is evident in the trail, because the hind tracks appear

The 3 × 4 lope of a Fisher in a dusting of snow.

The 2 × 2 lope of an American Marten.

A hopping Snowshoe Hare.

A stotting Mule Deer.

parallel to each other. Any local park should present ample opportunities to study squirrels using these gaits. Hops are similar to lopes, in that there is one moment when the animal is airborne during each cycle of footfalls, just after the hind feet push off. Bounds parallel gallops, in that there are two times when the animal is airborne during each cycle of footfalls—first after the front feet push off, and again after the hind feet push off.

The difference between the track patterns of hops and bounds is found in the relationship between hind and front tracks. When hopping, an animal's front feet land in front of the hind feet. Hopping is less common than bounding but can be observed in large voles, Muskrats, flying squirrels, and toads and frogs.

Hopping and bounding begin in the same way: the front feet either land as a pair (next to each other), or one after the other (one in front of the other), but in bounds the hind feet move forward beyond and to either side of the front feet. The front feet pick up as the hind feet pass to the outside, and there is a moment where the animal loses contact with the ground before the hind feet come down and push off again. This push-off is followed by a second moment in the air, before the front feet touch down and the cycle begins again. Numerous species bound, including squirrels, chipmunks, and rabbits.

Stott or Pronk

An unusual gait used by Mule Deer, Pronghorn, Elk, and occasionally other mammals is the *stott*. In this bouncing gait, an animal pushes off

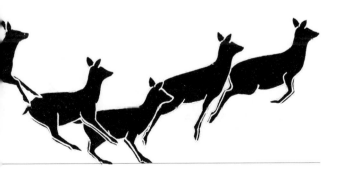

with all four feet at the same time, and then lands upon all four feet simultaneously, or nearly so. Mammals moving in this way appear to be using pogo sticks. Although not as fast as the gallop, the stott is better suited to traveling quickly over broken terrain, as well as allowing an animal to respond to external stimuli more quickly and to change direction when its feet hit the ground.

The trail of a stotting Mule Deer.

The bipedal hop of a kangaroo rat.

Bipedal Motion: Gaits on Two Legs

Bipedal Hopping

Both hopping and skipping are saltorial motions in which the front limbs remain elevated off the ground, and the resulting track patterns include paired hind footprints. Few mammals use bipedal motion, and in California the most likely culprit will be a kangaroo rat or kangaroo mouse. Birds, too, hop and skip. In hopping trails, paired hind tracks appear right next to each other, or nearly so, and the gait is the typical kangaroo-style hop. This pattern is possible because both hind feet hit the ground simultaneously. Technically, feet would only truly hit simultaneously if the feet were placed exactly next to each other, or the animal were coming straight down. However, for our purposes we'll use the word "simultaneously" to mean at the same time, or nearly so.

Bipedal Skipping

In skipping trails, tracks are also paired, but each hind foot lands completely independently of the other. Looking at the trail pattern, a hop becomes a skip when one hind foot registers completely in front of the other hind track. When kangaroo rats skip, they stay very close to the ground and take very long strides. Their hind feet rotate forward, one striking down before the other, and as the body moves forward over this foot, the second foot touches down. Their momentum continues, propelling the body forward over the second foot. As the body continues forward, the first foot to have touched down lifts up behind the animal. Continuing forward, the second foot joins the first behind the animal, then the animal lifts off, and together the hind feet rotate forward to begin another cycle. Momentum is more horizontal than vertical, and very little energy is wasted in rise.

Interpretation of Track Patterns

Until now we have discussed how an animal is moving in the technical sense of which limbs are moving when and in what order limbs make

contact with the ground. Next, we leap into speculative interpretation based upon our initial understanding of how an animal is moving, and ask ourselves why an animal is moving in a particular way. Because we can never know what an animal was thinking or exactly what an animal was doing at a particular moment (unless we actually witnessed the event), we speculate on its behavior and make our best guess. As in all aspects of tracking, we build a working hypothesis and test it as we continue to follow the animal. As we gather more information as we follow the animal, we either toss out our original hypothesis and create a new one, or continue to refine and support it.

Natural Rhythms and Energy Efficiency

Gaits and their associated track patterns are reflective of energy conservation, substrate, and behavior. Energy efficiency is a tremendously important variable in interpreting trails, as well as in predicting how a given animal will move in a given depth of substrate. It is certainly true that the slower an animal moves, the less energy it expends; however, it's more complicated than this. Every species possesses an anatomy, locomotion, and set of behaviors that improve its *fitness*. Fitness is quite simply the quality of success experienced by an animal, and in ecology, fitness is measured by how well an animal reproduces and perpetuates its genetic lineage. An animal has high fitness when it produces numerous offspring that themselves produce many offspring. Think of the measure of success as the number of grandchildren an animal has, because having grandchildren indicates successful production of offspring that were, in turn, able to survive and reproduce.

To simplify things further, let us focus upon only energy input and output. Each species balances its energy intake, meaning food collection, and energy output, which includes the ground covered to locate and/or pursue said food. Weasels have high metabolisms, insatiable appetites to meet their energy requirements, and gait preferences that propel them quickly over long distances. A weasel that walks too often may save energy in the short term but is denying that which makes it a weasel, and will die of starvation in the long term because it does not gather enough food to meet its energetic requirements. The balance point between energy intake and output influences the ways animals move, and the common gait that an animal has adapted to use for traveling is called its *natural rhythm* (Elbroch 2003). Let's compare Bobcats with Coyotes.

Each species has a body structure and biology that balances its individual energy intake and energy expenditure. Bobcats *walk*. Cats move through the forest slowly and stealthily, sitting and pausing frequently to study their environment, hoping to see or sense potential prey before they themselves are noticed. They may even lie in wait for prey and do not generally cover long distances while hunting. When a potential prey is selected, they stalk in and, when close enough, explode with enough speed to catch their intended victim before it is aware of them, or before it can

On the left is the trail of a trotting Coyote and the animal's resulting drag lines, and on the right, the clean walking trail of a Bobcat. Study the distance between footprints and the width of the trails in both animals.

escape. They grip their prey with their curved claws and deliver a killing bite to the head or neck.

Coyotes *trot* through the bush, hoping to catch smell, sound, or sight of potential prey, or to startle something to flee before them. Of course Coyotes do occasionally stalk, but in general, Coyotes cruise longer distances than medium-sized cats, allowing scents and sounds to betray the presence of prey species. When opportunities present themselves, they run down their intended prey, gripping and subduing it with their teeth. Their claws, which are never sheathed, project straight out from their toes and aid in traction while running.

Can Coyotes walk? Of course. Can Bobcats trot? Yes. Every animal is capable of a variety of gaits and speeds; however, each animal will have one or two gaits that are the most energy efficient for them, and they use these most of the time. Their common gaits are their *natural rhythm.*

Understanding the natural rhythms of animals is important, because it gives us a place to start. Think of an animal's natural rhythm as its normal speed, or middle ground. Each time an animal shifts from its normal speed, there is a reason and thus an opportunity for us to speculate why. Consider yourself as a starting point to better grasp how one might approach trail interpretation. You have your typical walking gait, speed, and subsequent track pattern. Should you be late for work or very focused upon a specific destination, your pace and track pattern will change. If you are hungry and stand within sight of five great restaurants, you will probably wander a bit and then move with determination once you have decided where to dine. You jump if you are scared and run when your life is threatened. The list of possibilities goes on and on. Why wouldn't all these sorts of changes be apparent in wild animal trails? The answer is that they are.

Certain gaits are associated with specific environmental conditions (substrates), while others betray behaviors or intentions. From a series of footprints you can determine whether an animal feels exposed and potentially uncomfortable, or whether it is hunting. And if you really become familiar with an area and its inhabitants, the way an animal behaves and moves may betray that it is out of its usual territory, or allow you to identify a transient or trespasser.

Interpretation of track patterns is advanced tracking and only comes with experience and lots of mistakes. The more natural history information you have about a species and the more time you spend trailing a particular animal, the better armed you are for this process. Regardless of experience, there will always be trails that will perplex you. Be reconciled with the fact that tracking is not a perfect science but a lifetime of learning.

Beginning Interpretation

With experience, interpretation can be highly detailed, and we discuss advanced aspects of interpretation later in the book. Here we encourage you to begin by interpreting the way an animal is moving as one of three categories: slow, normal (natural rhythm), or fast.

1. Slow—a speed slower than its normal gait (natural rhythm), that might indicate foraging, stalking away from danger, hunting, scent marking, exploring another animal's scent post, or numerous other potential behaviors.
2. Normal—its natural rhythm, the common gait in which an animal moves. The animal is exhibiting little or no undue stress, and is not blatantly reacting to any stimuli in the environment. You might say that the animal is acting "casually."
3. Fast—a speed faster than normal that might indicate that the animal is chasing prey, being chased by a predator, or startled by something in its environment. The animal may also be exposed and/or uncomfortable in their surroundings.

A portrait of a Desert Kangaroo Rat hiding in cover in the Mojave National Preserve in southern California.

Each of the three speed categories is associated with some general interpretations. The key to applying this to some animal you are tracking is in knowing what is their *normal* gait and associated natural rhythm. You can acquire this knowledge through experience in the field, watching animals in videos, or reading guides to tracking and animal behavior. As described above, the normal gait for Bobcats and Cougars is a walk, but the normal gait for Coyotes and foxes is a trot. The normal gait for squirrels is a bound, and for kangaroo rats, a bipedal hop.

Consider the Desert Kangaroo Rat (*Dipodymys deserti*), an animal the authors have never seen use any but three gaits and speeds: slow, normal, and fast. When moving slowly, Desert Kangaroo Rats *bound* on all four feet, like a squirrel, and they drag their tail on the ground behind them. They move in this way when they are foraging and scent marking, or when they are exploring the scents of other kangaroo rats. The normal gait of Desert Kangaroo Rats is a *bipedal hop*, which they use to travel between their burrows, foraging areas, and areas where they leave territorial scent marks. The bipedal hop is generally a sign of comfort and that everything is normal. Desert Kangaroo Rats *skip* to evade predators (it is a sign of fear) and in territorial chases with other kangaroo rats, including the courtship rituals that precede mating. Thus, we have three clear categories of movement, three speed categories, and three distinct categories of behavior associated with these gaits and track patterns. This of course can be done for any and every animal you follow.

In mammals that use more than three gaits, consider dividing each of the three categories into three; thus, you have nine potential speeds and interpretations. For example, Mountain Lions (or Cougars) use an overstep walk, or amble, as their normal gait. They speed up into a trot when exposed and crossing open ground, traveling to a known destination with some agenda in mind, and when confronting conspecifics in territorial encounters (stiff-legged trots communicate dominance in many

Bounding trails with tail drag of slowly moving Desert Kangaroo rats.

Rapid skips of an evading Desert Kangaroo Rat.

The beautiful trail of a large black bear sloshing in an overstep walk up a muddy riverbed in the Los Padres National Forest in southern California. Killdeer trails cross in the forefront of the photo.

mammals). They gallop when pursuing prey or startled, and lope when slowing down when they recognize that an initial threat is not as dangerous as they first thought, or is remaining stationary—three versions of "fast," each with different potential interpretations. But start with just three categories, and then add more as you feel more comfortable with the basics.

The Effects of Substrate on Gaits

Let us use ourselves as models. Compare the trails you leave while walking in an inch of snow and two feet of snow, or on firm ground and in deep,

In deep snow a black bear is forced to use a direct-register walk, as seen here in this bear switching dens in the middle of winter near Tahoe. Compare this trail to that on page 67, where a bear moves in shallow mud.

dry sand if you live in arid regions. It is likely that the length between your tracks decreases in deeper snow (softer, deeper sand) and that the width of the entire trail pattern increases. This is a wonderful lesson in the effects of the depth of substrate upon trail characteristics. Now, let's add another variable: speed.

When looking at trail characteristics in relation to depth of substrate, it may be more accurate to discuss *energy output* rather than *speed*. Let us return to our two trails in snow. If you were to use the same amount of energy in each trail, you would move more slowly in the deeper conditions. Moving at the same speed in both trails would require a higher degree of energy output to maintain that speed in two feet of snow. Consider running in the two snow conditions. Could you run at the same speed in two feet of snow as in an inch of snow? Would the energy output be equivalent while maintaining a run in these two very different conditions?

Just as we adapt to changing conditions, so do other animals. In fact, as you track across varied substrates, you will begin to note that animals change the way they are moving in ways that reflect an awareness of energy efficiency. Trotting in deep snow is intensively difficult, if not impossible at times, and so Coyotes are often found walking or bounding for short distances in these conditions. Fishers, which tend toward a 3 × 4, or rotary lope, change tendencies to a 2 × 2 or transverse lope in deep, soft conditions, and often walk longer distances as well.

Tracking in Snow

Snow tracking varies from a simple exercise in perfect track identification to incredibly difficult interpretations of deep, windblown trails. Below are some techniques and approaches useful when interpreting tracks and trails in snow:

TRACK PATTERNS. Learn to rely more on track pattern identification than print identification, and in this way be able to identify species from greater distances and through binoculars. In open areas of the North, biologists complete track transects of far-ranging mammal species, such as Wolverines, by small plane.

GAIT CHANGES. Be aware of the gait changes animals make when moving in snow. Animals that do not typically direct-register walk suddenly begin to do so. As an example, consider the raccoon, which uses a distinctive 2 × 2 walking gait in shallow substrates. But in snow, raccoons use a direct-register walk. Also mammals that tend to trot or lope in shallow substrates, walk more frequently in deep snow. Remember that the depth of substrate is one of the key variables in determining how an animal moves.

DETERMINING DIRECTION. The direction of travel in a given trail can be determined by identifying the deepest part of an individual track (the deep end points forward), but check several tracks to be sure. Place your hand into trails filled with snow to feel the direction of travel. This is best done without gloves, so be prepared and be conscious of conditions.

TOUCHING TRACKS. Feeling snowed-in tracks offers more than the direction of travel. The mound in the center of canine tracks can often be clearly felt, or the ridge between palm and toe pads in a cat track, or hooves. You'll also be able to tell the approximate size of the track, and how flat the floor of the track is. For example, the floor of canid tracks in deep snow is significantly steeper than the floor of tracks made by felids.

Feel snowed-in tracks very gently without gloves. The area compressed by the animal will always be firmer than the snow that has blown or fallen in. Use the body heat of your hands to melt out tracks; in this way you can sometimes not only feel the track but recreate it visually. However, do not force fingers into spots too quickly so as to create the track you want; carefully and slowly melt out the existing track floor.

BLOWING OUT SNOW. When light snow has filled in a set of tracks, you can often blow them out and keep the track intact. This is an especially useful technique when temperatures are cold and a fresh, light layer of snow has covered tracks.

CLAWS. Look for the placement of claws in snow tracks. In deep snow, all that may be clear in Coyote tracks are the pinprick claw marks of toes 3 and 4 at the very end of the track; you may have to squat to see them. Tracks of Bobcats and other cats may be differentiated from canine tracks by looking for claws that register in the snow on a higher plane than the

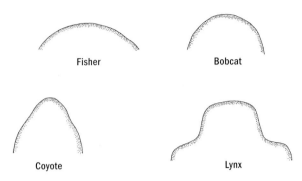

| Fisher | Bobcat |
| Coyote | Lynx |

In deep snow, when track details are obscured, the shape of the front wall created by the foot as it enters the snow is especially useful in species identification. Compare the shapes of the walls for the species presented here.

floor of the track. The claws will appear to be above the toes, because this is where they sit when "sheathed."

TRACK WALLS. The walls of tracks created as a foot enters and exits snow are filled with clues for species identification. The back wall found behind the track may hold a metacarpal pad or dew claws on a higher plane than the track. The shape of the front wall is especially important for interpretation. Is the track pointed or rounded? Refer to the illustrations for examples.

DRAG MARKS. The drag marks found between the tracks are also useful. One method in which gray squirrel and cottontail trails can be differentiated is by studying the drag marks made between sets of four tracks; squirrels tend to drag on the outside edges of the track pattern, while cottontails tend to drag along the median line.

The shapes of the drag marks and the entrance/exit holes along the surface of the snow are also worthy of study. As an example, consider the shapes left by Bobcats and Domestic Cats when walking in snow. They move in such a way as to create perfect triangles. Or look for the paired lines created by dragging hooves in deer trails.

Recognizing Track Patterns for Interpretation

There are also specific track patterns with which you will want to become familiar, so you can quickly interpret them in the field. With practice you will be quickly able to surmise what an animal was doing from a peculiar series of tracks—sniffing something, pausing, urinating, feeding, etc. Deer "point" to food with a single hoof, many mammals "T-up" when they pause (see later), and Bobcats kick out a hind leg when passing an object they intend to spray with urine. Study the accompanying illustrations to help get you started.

A feral hog demonstrates pointing; look for the track angling off to the left, and just beyond, the mark of the animal's snout where it rooted and devoured some vegetation in passing.

Pointing

Follow a deer trail in good substrate for a short distance, and you will probably find a spot where a front track registers off to the side, an anomaly in an otherwise redundant zigzagging pattern of tracks. This is a front track that was positioned to hold the weight of the head as it reached out and browsed passing vegetation. The leading edge of deer tracks is pointed, and the "arrow" created by this track often points directly to the vegetation browsed by the deer. Pointing can be found on all the ungulate trails, as well as where carnivores step out to investigate a passing scent.

Sit Downs

Learn the shape of patterns created by sitting mammals. "Sit downs" are a feature of hunting Bobcat and Cougar trails, which often pause where views of prey are good or in prey-abundant areas. If they wait for a longer period, they are more likely to lie sphinxlike rather than on their sides, as when resting. Canids sometimes sit in areas with wonderful views or when soliciting a reaction from a conspecific. Sitting in both felids and canids is also a sign of curiosity and investigation, because they often sit to watch something with which they are unfamiliar. Sitting is also common at den and rendezvous sites, and near kill sites of larger prey.

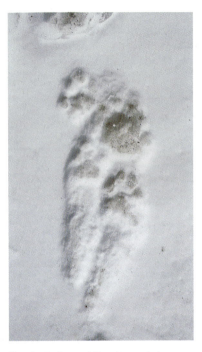

The sit of a Cougar kitten demonstrates the complete hind feet, including the heels, so rarely seen in many moving digitigrade mammals.

A Bobcat in Santa Barbara County illustrates the "T stop," as it pauses on a hunting round. The "T" was created when the Bobcat broke its walking pattern and placed its left front foot on the ground adjacent to its right front.

"T-Trails" and "Box Stops"

When walking and trotting animals pause momentarily or stop for longer periods, there is often a "T" in the trail. The vertical line of the "T" shape is the typical trail pattern, and the horizontal slash (the "cross" on the T) is created by two front tracks sitting next to each other that break the typical rhythm of footfalls. T-trails are common where animals have heard something in the distance, are pausing to investigate a road before crossing, and have paused at trail junctions. On rare occasions the horizontal slash in the T can be made by two hind feet instead of two front feet; in these cases the animal holds a front foot up and stabilizes itself with the remaining three legs.

A variation of the T-trail is the box stop, where an animal stops and places all four feet on the ground at the same time. Typically the two corners of the "box" that are part of the normal trail pattern are deeper and

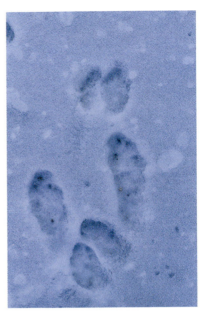

A Ringtail in the Sacramento Valley demonstrates the "box stop." Look carefully to see the lighter prints of the left front and right hind feet.

The double fronts of a Snowshoe Hare where it paused—one set behind the larger hind tracks, and the second set in front of them.

easier to see; the two additional "corners" are the extra front and hind tracks that fall outside the normal rhythm of footfalls, and these tracks are often lighter and harder to see.

Double Fronts

A pause or stop in a bounding animal such as a cottontail or squirrel is obvious in a trail section where one pair of rear tracks is accompanied by two sets of front tracks. When the animal is bounding normally the front tracks touch down and are followed by the rear tracks beyond them. When it is time to stop, the rear tracks stay put, but the front feet, which have just picked up to allow the rear feet to register beyond them, touch down a second time in front of the rear feet to stop forward momentum. When it is time to move on, the front feet are lifted up and the rear feet push off.

Determining the Position of the Head

The position of an animal's head can often be determined by looking for the deepest part of an individual front track; however, large head movements are often more easily determined by studying the overall track

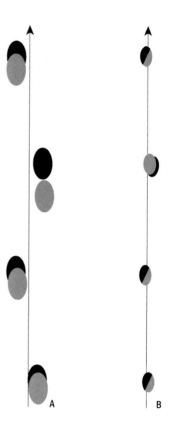

In both trails the animal is looking to the right. In trail A the animal is walking and turns more to the right, and the turn occurs over a longer duration of time. In trail B the animal is trotting; the look to the right is much shorter and the head is not turned as far.

A

B

pattern. What we describe here is a method to determine the position of the head, and potentially where it was looking or smelling, that works for the slower gaits of walking and trotting. When walking, an animal that shifts to look over its right shoulder will typically influence three pairs of tracks and cause an understep on the side of the body to which it turned. In contrast, when an animal is trotting, the same maneuver often causes a slight overstep, because the momentum of the hind feet continues even though the front foot lands short on account of the shift in the weight of the head. Simple role playing in sand or other suitable substrate will prove out these concepts.

As a footnote, there has been discussion in the field about the usefulness of interpreting where a canid is looking by which side of the animal is kicked out during a side trot. However, canids can look wherever they want, regardless of which side the rear end is angled.

Measuring Mammal Tracks and Trails

A ruler is a wonderful tool to help you build confidence in your perceptive and intuitive skills. Track and trail measurements aid in identification and are critical for scientific documentation of species in your area; when you are documenting a rare or potentially controversial species, *always* take measurements and include a ruler or other scale in photographs. Measurements can also be shared in research and used as a means with which to compare species characteristics. Tape measures, however, if relied upon for too long to aid you in species identification, will likely become a hindrance to further developing your perceptive and intuitive skills. For instance, the registration of toe 1 in the front tracks of the Douglas's Squirrel is a less ambiguous, and therefore more useful, means with which to differentiate their tracks from those of other tree squirrels. Learn to look for and appreciate the details.

Measuring Tracks

Here track parameters include nails and posterior pads. Some trackers do not include the nails; however, certain mammal species, including porcupines and pocket mice, rarely register digital pads, and thus measurements would be impossible without them. Also, be cautious to avoid measuring drag marks or the impressions made by feet as they enter and leave a particular substrate; measure the floor of the actual track and not the walls.

There are several exceptions. All hooves are measured *without* the dewclaws, because they register inconsistently in many species and we want you to be able to compare numbers across species. Carpal pads, which lie high on the legs of canines and felines, are not included in measurements, because they only register at high speeds or in deep substrates; they are, however, included within the parameters for mustelids and rodents, because they register often.

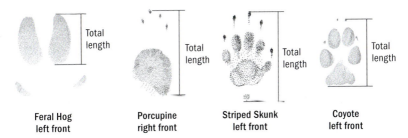

| Feral Hog | Porcupine | Striped Skunk | Coyote |
| left front | right front | left front | left front |

Four versions of total length (TL): include the claws and all pads, except in ungulates where the dewclaws are not included in the measurements. Essentially, measure what you see; sometimes pads will be lacking, and your length will be shorter.

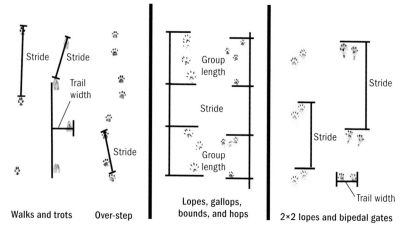

| Walks and trots | Over-step | Lopes, gallops, bounds, and hops | 2×2 lopes and bipedal gates |

How to measure trails.

Always measure several front and rear tracks of any species you encounter. An individual animal can show incredible variation in track size, and within a given mammal species there is an even larger variation. For this reason, this book provides track parameters, rather than averages, for easier identification. There is often concern about making parameters too large and thus less useful. We support the practice of parameters, even large ones. We would rather that you consider more potential species in your interpretation than ignore the correct interpretation because you measured an unusual splayed footprint.

Measuring Trails

Walks and trots are measured in the same way, from the tip of the front track to the tip of the very next front track (on the opposing side of the body). Direct-registering trails are measured from the tip of the rear track to the tip of the very next rear track. Straddle and side trots are also measured from front track to front track. When parameters in walking trails created by measuring from rear to rear were compared with those made by measuring front to front, rear-to-rear parameters were very similar yet always larger. The trail widths of walks and trots are taken at their widest points. Additional measurements can be useful comparisons across species, such as the group length of paired tracks in a side-trot trail.

Lopes, gallops, bounds, and hops are measured in the same way, with the exception of the 2 × 2 gallop used by weasels. In all these gaits, the length of each group of four tracks is measured as the group length. The stride is the distance in between groups of four tracks.

The 2 × 2 gallop is unique. Although it is inconsistent with the other lopes and gallops, we were reluctant to throw out years of data collection just for consistency's sake. Measurements were taken from the tip of the lead track in a pair to the tip of the lead track in the very next pair. The trail width is also measured straight across—not at an angle—and is a useful tool for species comparisons in weasels.

The trail width is also a useful measurement in bounds and hops, and is taken at the widest point of the trail—the distance between and including the two hind tracks.

Mammals Species Accounts

This book is primarily about California's mammals. Starting on page 121 are species accounts of track and trail information for most mammals found in the state. In addition, a quick reference guide is available on plates 13 to 84, which begin after page 98; numerous life-size illustrations should aid you in quick identification in the field. All drawings are made from slides of footprints taken in natural settings, and were drawn with painstaking accuracy to convey real tracks in real substrates.

TRACKS AND TRAILS OF BIRDS
AND OTHER ANIMALS

Mark Elbroch

The left track of a Great
Horned Owl.

Countless tracks and trails you will encounter in California will not be
made by mammals but by other animals. Here we attempt to better prepare you to begin to interpret the signs of numerous birds, as well as some
other common creatures. This training is complementary to that for
mammal tracking, because the tracks of all animals can be confused with
one another, and if you catch the tracking bug (as each of the authors
has), you will want to know what made every mark in the ground you
encounter. There seem to be endless tracks and trails that still need study
and documentation, but given our space limitations, this section should
get you started.

Birds

Foot Morphology

The structure of bird feet and the arrangement of their toes aid ornithologists in the organization and classification of species. A bird's foot structure also tells you something about how a bird exists in the world—its
foraging and nesting habits—and so we use foot structure as the basis
for presenting bird tracks in easily learned groups based on common
characteristics.

Most birds are four-toed; a few are three-toed. The toes of birds all
converge in a central area of the foot called the metatarsal pad, or sole.

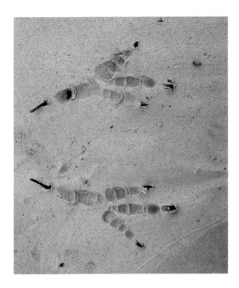

Common Raven tracks in Death Valley National Park exhibit the typical anisodactyl foot structure.

Birds do not place all the bones of their feet equally on the ground; birds are digitigrade and usually move with more weight supported by their toes, creating deeper impressions at the outer edges of their tracks instead of in the center.

Bird feet vary tremendously depending upon how the toes, or digits, are arranged around the metatarsal pad. The digital pads that line each toe also vary in form and function from bird to bird. The most common toe arrangement is anisodactyl—with three toes pointing forward and one backward—creating the foot structure of nearly every cartoon bird you've ever seen. The toe that points backward is commonly referred to as the hallux. The hallux is particularly useful in differentiating among bird tracks, because it varies in length across species and is absent altogether in some.

Ornithologists number the toes of each foot, from 1 to 4, in a systematic fashion, called the digital formula. Toe 1 is always the hallux, whether it is present or not, and the other toes are numbered in sequence, beginning with the inside of the foot and circling out. Thus, left and right tracks are circled in different directions (see the figure on page 82). For anisodactyl feet, toe 1 points backward and toes 2, 3, and 4 point forward. The second most common toe arrangement is zygodactyl, with two toes pointed backward and two forward. For zygodactyl feet, toes 1 and 4 point backward and toes 2 and 3 point forward.

Track Categories

Most bird tracks fall into one of five categories depending upon their foot morphology and resulting visual characteristics. A few birds fall

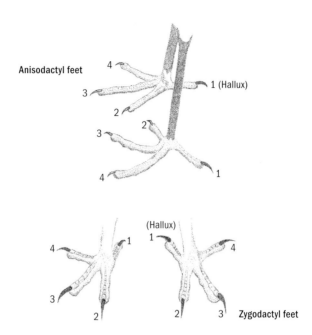

Anisodactyl feet

4

3

2

1 (Hallux)

2

3

2

1

4

(Hallux)

1

1

4

4

3

2

2

3

Zygodactyl feet

The digital formula for bird feet.

outside these categories, which actually facilitates their identification. These include swifts, rarely encountered, and three-toed woodpeckers.

Classic Bird Tracks

Classic bird tracks are those technically termed anisodactyl. This is the bird track most people would draw without any prompting: three toes pointing forward and one pointing backward. It's toe 1 that points backward and toes 2, 3, and 4 that point forward. Many bird species share this track category. Kingfishers are technically syndactyl, with toes 2 and 3 fused for a portion of their lengths, but they still leave classic bird tracks.

Measuring classic bird tracks is quite simple. Always include the claws, and measure the track length from the tip of the claw of toe 1 to that of toe 3. The length of bird tracks is less variable than their width, and therefore more useful.

Bird families that fall into this category include bitterns, herons, egrets, ibises, vultures, eagles, hawks, falcons, moorhens, doves, kingfishers, flycatchers, shrikes, vireos, jays, nutcrackers, magpies, crows, ravens, larks, swallows, titmice, creepers, nuthatches, wrens, dippers, kinglets, gnatcatchers, thrushes, mimics, thrashers, starlings, pipits, waxwings, warblers, tanagers, towhees, sparrows, longspurs, buntings, cardinals, bobolinks, meadowlarks, blackbirds, grackles, cowbirds, orioles, and finches.

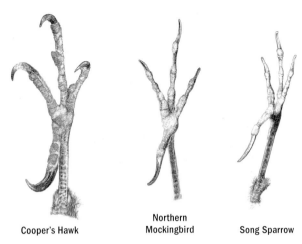

Cooper's Hawk   Northern Mockingbird   Song Sparrow

Three examples of classic anisodactyl foot structures.

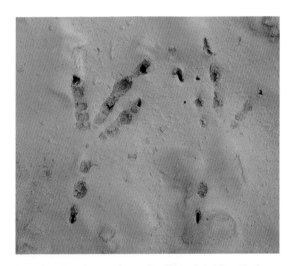

Clean American Crow tracks can be differentiated from the larger tracks of ravens by studying the shape of the digital pad closest to the metatarsal on the hallux. In crows it is the same width as the remainder of the toe, and in ravens it is trapezoidal and wider than the toe.

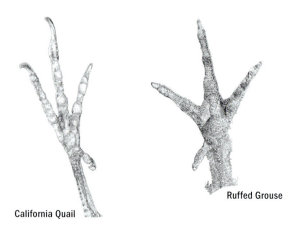

California Quail

Ruffed Grouse

Two examples of game bird tracks, which are an anisodactyl structure with reduced or absent halluces.

California Quail tracks exhibit the "game bird" morphology.

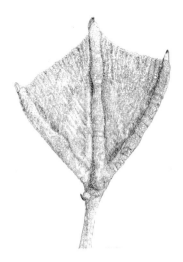

This Western Gull illustrates a palmate foot structure.

Game Bird Tracks

In game bird tracks (plates 50–57), toe 1, the hallux, is significantly reduced or absent altogether. In many species, toe 1 is physically located above the metatarsal pad and remaining toes. A foot structure in which the hallux is elevated is termed incumbent. Technically, game bird tracks are anisodactyl, in that the skeletal nub of toe 1 points backward and toes 2, 3, and 4 point forward; however, we differentiate them based on the structure of toe 1. In addition to game birds such as grouse and quail, shorebirds also fall in this category.

Measure the length of game bird tracks from the tip of the claw of toe 3 to the far end of the metatarsal pad. Because toe 1 does not reliably show up in tracks of game birds and shorebirds, we don't include them in measurements. This allows us to compare measurements across species, even with those that do not show a hallux at all.

Families in this category include pheasants, turkeys, grouse, ptarmigan, quail, rails, coots, cranes, plovers, oystercatchers, stilts, sandpipers, and phalaropes.

Palmate (or Webbed) Tracks

Palmate refers to the distal (full) webbing between toes 2 and 3, and between toes 3 and 4. Palmate tracks (plates 58–62) are similar to game bird tracks in that toe 1 is greatly reduced, incumbent, and/or absent altogether; however, they are also distinctive because of their webbing. Like tracks of game birds, palmate tracks are technically anisodactyl.

Measurements are taken exactly as for game bird tracks: toe 1 is not included in the length, because it shows unreliably in the tracks of some species. Families include loons, swans, geese, ducks, avocets, gulls, and terns.

A palmate Canada Goose track and accompanying tracks of a Spotted Sandpiper.

The totipalmate structure of a Brown Pelican foot.

Totipalmate Tracks

Like palmate tracks, totipalmate tracks (plate 63) exhibit distal webbing between toes 2 and 3 and toes 3 and 4. But totipalmate tracks also exhibit webbing between toes 1 and 2; every toe on the foot is webbed. Should the webbing be unclear, totipalmate tracks can also be distinguished because toe 4 is the longest toe in the track, significantly longer than toe 2 and noticeably longer than toe 3. In palmate tracks, toes 2 and 4 are relatively similar in length, and toe 3 is the longest toe in the track.

The zygodactyl foot structure of a
Great Horned Owl.

American White Pelican tracks, an example of
totipalmate foot structure.

Measure the full length of totipalmate tracks from the tip of the claw
of toe 1 to the tip of the claw of toe 4. Families include boobies, gannets,
pelicans, and cormorants.

Zygodactyl Tracks

In zygodactyl tracks, toes 1 and 4 point backward, and toes 2 and 3 point
forward. Ospreys, woodpeckers, cuckoos (which include roadrunners),
and parrots consistently show this pattern clearly. Owls also fall into this
group, but because they can rotate toe 4 into different positions, their
tracks can potentially show great variability. However, owl tracks in the
field tend to conform to a consistent shape. Instead of two toes forward
and two backward, owl tracks often have two toes forward, one sideways,
and one backward.

Measure the length of zygodactyl tracks at their longest point. Do not
measure angles, which increases the measurement, but straight up the
length of the track. For owls, measure along the inside of the track, from
the tip of the claw of toe 1 to the tip of the claw of toe 2.

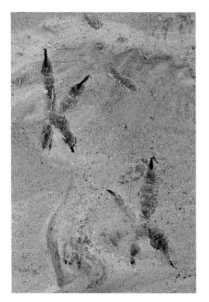

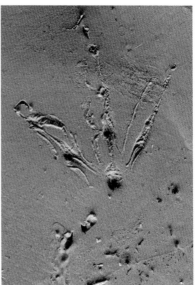

The zygodactyl tracks of a roadrunner.

The abnormally large and distinctive track of an American Coot near San Francisco.

Studying Bird Tracks in the Field

Useful concepts we use in the identification of bird tracks are relative size, symmetry, the shape of the digital pads, metatarsal registration, and webbing. Let's take each in turn.

Relative Size

The relative size of a track provides useful information for guessing or extrapolating the probable size of a bird; however, sometimes track size can be misleading. Species that spend a great deal of time on the ground have developed larger feet in proportion to the size of their bodies. For example American Coots have massive feet for their size, and kingfishers tiny ones. In addition, different species have evolved larger or smaller feet to better exploit specific ecological niches, such as for swimming, gripping prey, or hanging upside down.

Symmetry

Symmetry is the comparison of the left and right sides of the track. If you were to cut a track lengthwise in half by following the line right down the middle of toe 3, the symmetry would describe the degree to which the left and right sides match. Bird tracks are rarely perfectly symmetrical, but the degree of asymmetry is useful to note and compare across species.

Compare the bulbous digital pads in the Bald Eagle tracks pictured here and the Great Blue Heron tracks in the accompanying photo.

A pair of Great Blue Heron tracks.

Shape of the Digital Pads

Consider the tracks of Bald Eagles and Great Blue Herons. Even though simple measurements of the length and width of their tracks are similar, there are obvious differences in the structure, or morphology, of their feet. The massive claws and robust, rounded, rough toe pads of the eagle aid it in gripping slippery prey. The slender, wide, flat feet of the heron, and its skinny toes lacking any distinct digital pads, are ideal for supporting it during long waits in muddy marshes.

Presence of the Metatarsal Pads in the Track

All birds have metatarsal pads, but in many species they never register in the tracks at all. Thus, noting the presence or absence of a metatarsal pad is a useful starting point in your analysis.

Webbing

Visible webbing in tracks varies tremendously across species. In ducks and gulls, webs are large and clearly visible. In herons, grouse, and certain plovers, smaller webs are present between two or more toes. Because partial webbing does not always show in the track, species with partial webbing (termed semipalmate) are not grouped here with species that display obvious webbed tracks but with those whose tracks do not show webbing.

The beautiful walking trail of a Killdeer. Look for the semipalmate webbing deep between the toes, and note that a hallux has not registered at all.

Even the fully webbed feet of geese and gulls cannot be relied upon to leave clear webbing marks in their tracks. The depth and hardness of the substrate in which a bird stands affects a track significantly. In cases where webbing is not clearly visible, it can be deduced by noting whether the outer toes are curved. On webbed feet, toes 2 and 4 always curve in toward toe 3.

Gaits and Track Patterns

A gait describes the method of locomotion and the distinct way the body moves to propel a bird forward on the ground. There is a relationship between the gait of the bird and the track pattern left behind. Most bird track patterns show one of four distinct gaits: walking, running, hopping, or skipping. Individual bird species, like mammals, tend to use one particular gait more often than the others; this is their natural rhythm. However, a species may use a second and sometimes even a third gait as well. If you see the trail of a bird that is moving in a gait that is not its typical natural rhythm, ask yourself why. This will help you learn to interpret bird behavior from their trails.

Like in mammals, the gaits that birds use are especially useful in species identification. The running patterns of traveling thrashers make their trails impossible to confuse with the hopping trails of towhees sharing the same stretch of dirt. As another example, know that the only jay that walks is the Pinyon Jay. Because gaits are so useful in identification, we list those used by each species under their track lengths in the track plates that follow; gaits are listed from most commonly observed in the field to least.

Bird gaits are naturally bipedal; birds only have two feet to use. Having only two feet limits considerably the potential variation in how birds move as compared to animals that have more feet. Some bird gaits are identical to those described for bipedal motion in mammals, as you will see below.

The crisp trail of a walking male Brewer's Blackbird. If you look carefully you'll see the much smaller tracks of a female crossing his trail.

Bipedal Walking

Bipedal walking is the preferred gait for many birds that spend a great deal of time on the ground. All birds with a reduced or absent toe 1 walk. These include ducks, many wading birds, gulls, grouse, and quail. Most raptors walk when they're on the ground, as do crows, ravens, pipits, and magpies.

Walking is a slow gait in which one foot is placed in front of the other, carrying the bird forward in exactly the same way as we walk around. At least one foot is on the ground at all times, and once in each cycle both feet are in contact with the ground simultaneously. The remaining track pattern is of single tracks at regular intervals, in a zigzag fashion or straight line, depending on the width of the trail. In some species, such as quail, the straddle is narrow, because they place one foot almost directly in front of the other. In other species, such as doves or ducks, the straddle is wider, because they do not place each foot in front of the other.

The running trail and abrupt stop of an American Robin.

Bipedal Running

Birds run just like we do, and running is a gait used by birds that spend a lot of time on the ground. One foot is placed in front of the other in rapid succession, and there is a moment in each cycle where the bird is airborne and no foot is in contact with the ground. Running birds are easy to observe, especially in backyards with American Robins or along beaches with Sanderlings.

Similar to walking trails, running trails are created by single tracks left at regular intervals. The strides of running towhees are two and a half to three times their track length. Robins are similar. The strides of running magpies are four to six times their track length, and thrashers three to five times. Additional clues can be used to help identify a particular track pattern as a running gait. Are the tracks fully registering? Or did the bird shift its weight forward? Is there a greater disturbance or debris thrown back around the track? Has the track become blurred by the force and speed with which it was created? Has the track splayed more than usual? Has the trail width narrowed or become more pigeon-toed? These are all potential indicators that your bird is running.

Running trails are measured just like walking trails. Measure a stride from the tip of the claw of toe 3 in one track to that of toe 3 of the next track.

Hopping

Hopping and skipping gaits in birds are exactly the same as those described for mammals, and we encourage you to review them now. Both hopping and skipping are trails where tracks are paired instead of falling independently at regular intervals, as in walks and runs. In hopping trails, paired tracks appear right next to each other, or nearly so. This pattern is possible because both feet hit the ground at the same time. Juncos and finches provide wonderful examples of hopping. When you see birds hopping, note the vertical component of the movement, which creates an arcing line of travel between the moments when the bird is on the ground.

The paired tracks of a hopping Western Scrub-Jay.

Skipping

In skipping trails, tracks are also paired, but each foot lands independently of the other: A trail pattern is a skip rather than a hop when one foot registers completely in front of the other. Birds that skip tend to be small perching birds that nevertheless spend a great deal of time on the ground, such as Snow Buntings and Song Sparrows. One theory suggests that placing the feet in this way, rather than right next to each other, increases maneuverability.

When you see a bird skipping, it stays very low to the ground, which can give the impression of running. The head of the bird remains nearly level, and all the rotation comes from the lower body and legs. Both feet rotate forward, one foot strikes down, and as the body moves forward over this foot, the second foot touches down. The momentum continues propelling the body forward over the second foot, while the first foot lifts up behind the bird. Continuing forward, the second foot joins the first behind the bird, lifts off, and together they rotate forward for another cycle. Momentum is horizontal, and little energy is wasted with vertical rise.

Other Animals

The numbers of other wildlife in California far outnumber the birds and mammals, yet here they are provided only the briefest of introductions. Certainly the larger signs of warm-blooded vertebrate animals are easier to see and are often more engaging for the general public. Yet the frontier that is the tracks and trails of reptiles, amphibians, and invertebrates remains largely unexplored and is an easy area for someone with enthusiasm and determination to contribute to our growing field of wildlife tracking.

Reptiles and Amphibians

The feet of reptiles may be rough and pebbly or scaled, and all have well-developed claws; in contrast, the feet of many amphibians are smooth as glass and lack claws (see the figure on page 95).

The stunning fresh trail of a Pacific Green Sea Turtle shuffling back to the safety of the ocean after laying her eggs on a remote beach.

Look closely to watch the foot cycle in a walking salamander.

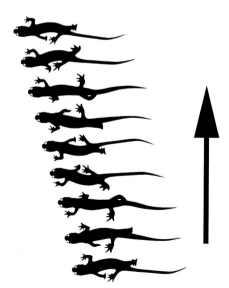

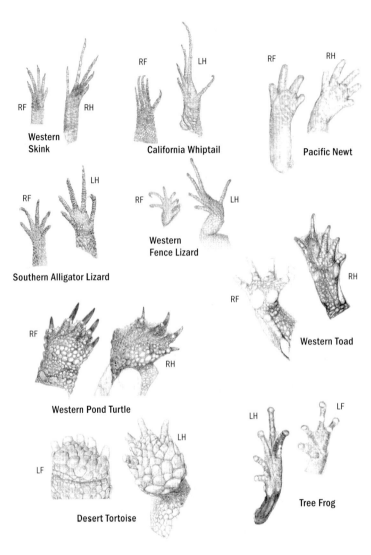

Feet of several California reptiles.

Reptiles and amphibians move on four legs, like mammals. Many of them move in exactly the same way as described for walking mammals, or at least similarly enough to begin to visualize them in action. Like bears and beavers, tortoises walk, one leg at a time. They do this to take advantage of the strength of the tripod configuration created when three feet are

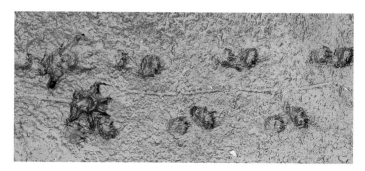

Paired lines of tracks made by a walking pond turtle, and several tracks of an opossum crossing its trail.

on the ground. They move too slowly in relation to their weight to move in any other way. This is useful to note: it is impossible for a turtle's hind foot to step over or beyond the front foot on the same side of the body, because the front foot is still on the ground when the hind foot comes up from behind. Thus, they cannot make direct-register or overstep track patterns, both of which are commonly made by mammals.

Numerous reptiles and amphibians, including lizards and salamanders, move opposing limbs—the right front and left hind, simultaneously, as if joined by a cable. This is a parallel movement to trotting, as described for mammals. However, unlike mammals, the feet of reptiles and amphibians do not move beneath them when they pull them forward to take their next step; instead they swing wide to either side. Most lizards and salamanders also have short limbs in relation to the length of their bodies, and so must flex their spines considerably to increase the length of their steps. This means that as a salamander stretches forth with its right front foot, its shoulder is angled one way, and that its hip is angled in the opposite direction. The greatest difference between salamanders and lizards is that salamanders walk slowly, never losing contact with the ground, and lizards often run (moving in a trot as described for mammals), during which they push off with one pair of opposing limbs and are airborne before touching down with the other pair of opposing limbs. At high speed, many lizards run bipedally on only their hind feet (Irschick and Jayne 1999).

Life-sized illustrations of a select few of California's reptiles and amphibians can be found on plate 67 to 73.

Invertebrates

The diversity in shape and species of invertebrates is mind-boggling, and thus the diversity of ways in which they move is equally varied. In six-legged species, they typically move opposing limbs together in groups of threes in the following fashion (somewhat like a six-legged trot or

The trail of a "trotting" lizard.

diagonal walk, where the legs are split into two supporting tripods): group 1 (right front foot, left middle foot, and right hind foot) move simultaneously together, and group 2 (composed of left front, right middle, and left hind feet) move together at alternate intervals. Some species, including cockroaches, only run on their hind four legs at top speeds (Full and Tu 1991). In some species the abdomen drags and in others it does not; in some species it only drags some of the time.

Life-sized illustrations for a select few species can be found on plates 74 to 84.

QUICK GUIDE TO TRACKS

Mammal Tracks

Bird Tracks

Reptile and Amphibian Trails

Invertebrate Trails

PLATE 13 Mammal Tracks: Actual Size

Shrew (*Sorex* and
Notiosorex spp.),
pages 144–146

Western Harvest Mouse
(*Reithrodontomys
megalotis*),
pages 343–344

Mole (*Scapanus* spp.),
pages 147–151

Red-backed vole
(*Myodes* spp.),
pages 326–329

Pocket mouse (*Perognathus*
and *Chaetodipus* spp.),
pages 308–310

California Vole
(*Microtus californicus*),
pages 326–329

Deer mouse (*Peromyscus*
spp.), or White-footed
Mouse (*P. leucopus*),
pages 339–342

Ermine, or Short-tailed
Weasel (*Mustela erminea*),
pages 222–227

Small and medium
kangaroo rat
(*Dipodomys* spp.),
pages 311–314

Jumping mouse
(*Zapus* spp.),
pages 323–325

Chipmunk
(*Tamias* spp.),
pages 301–304

PLATE 14 Mammal Tracks: Actual Size

Small woodrat
(*Neotoma* spp.),
pages 334–339

Roof Rat, or Black Rat
(*Rattus rattus*),
pages 347–351

Brown Rat, or Norway
Rat (*Rattus norvegicus*),
pages 347–351

Large woodrat (*Neotoma* spp.),
pages 334–339

Pocket gophers (*Thomomys*
spp.), pages 319–322

White-tailed Antelope Squirrel
(*Ammospermophilus leucurus*),
pages 286–287

PLATE 15 Mammal Tracks: Actual Size

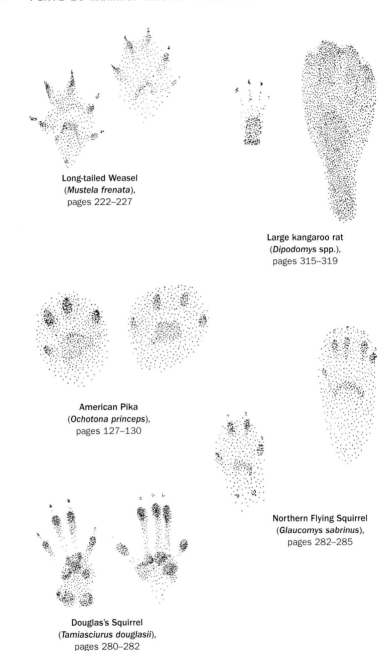

Long-tailed Weasel
(*Mustela frenata*),
pages 222–227

Large kangaroo rat
(*Dipodomys* spp.),
pages 315–319

American Pika
(*Ochotona princeps*),
pages 127–130

Northern Flying Squirrel
(*Glaucomys sabrinus*),
pages 282–285

Douglas's Squirrel
(*Tamiasciurus douglasii*),
pages 280–282

PLATE 16 Mammal Tracks: Actual Size

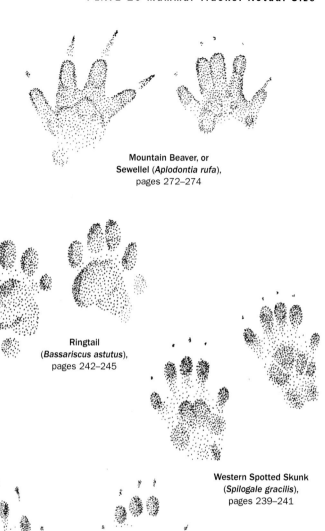

Mountain Beaver, or
Sewellel (*Aplodontia rufa*),
pages 272–274

Ringtail
(*Bassariscus astutus*),
pages 242–245

Western Spotted Skunk
(*Spilogale gracilis*),
pages 239–241

Medium ground squirrel
(*Spermophilus* spp.),
pages 296–298

PLATE 17 Mammal Tracks: Actual Size

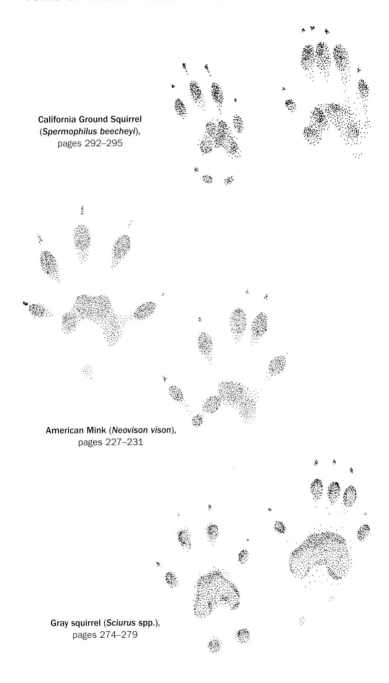

California Ground Squirrel
(*Spermophilus beecheyi*),
pages 292–295

American Mink (*Neovison vison*),
pages 227–231

Gray squirrel (*Sciurus* spp.),
pages 274–279

PLATE 18 Mammal Tracks: Actual Size

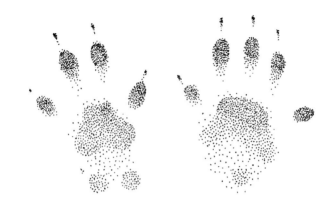

Eastern Fox Squirrel (*Sciurus niger*),
pages 274–279

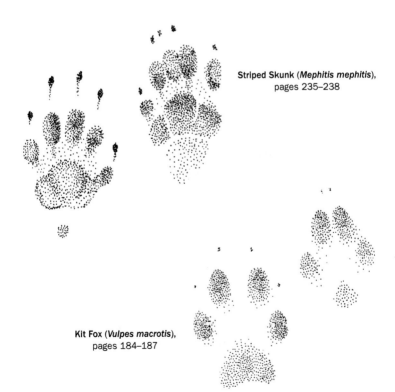

Striped Skunk (*Mephitis mephitis*),
pages 235–238

Kit Fox (*Vulpes macrotis*),
pages 184–187

PLATE 19 Mammal Tracks: Actual Size

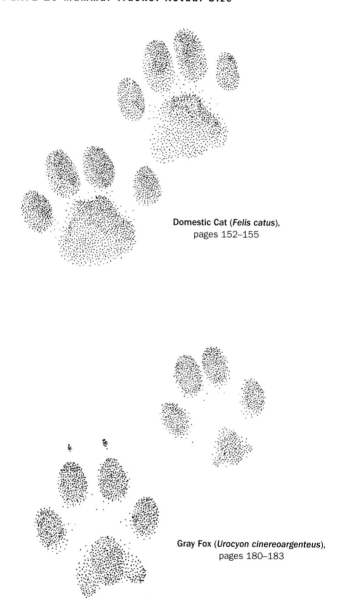

Domestic Cat (*Felis catus*),
pages 152–155

Gray Fox (*Urocyon cinereoargenteus*),
pages 180–183

PLATE 20 Mammal Tracks: Actual Size

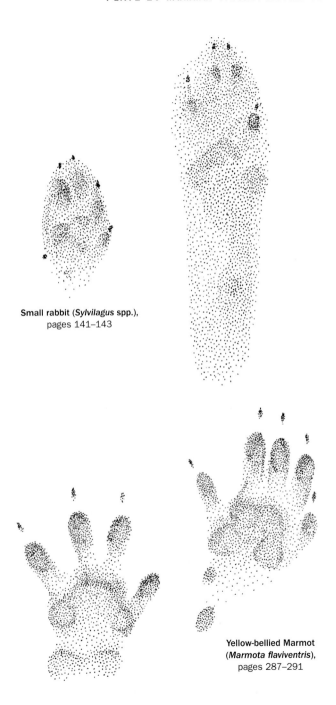

Small rabbit (*Sylvilagus* spp.),
pages 141–143

Yellow-bellied Marmot
(*Marmota flaviventris*),
pages 287–291

PLATE 21 Mammal Tracks: Actual Size

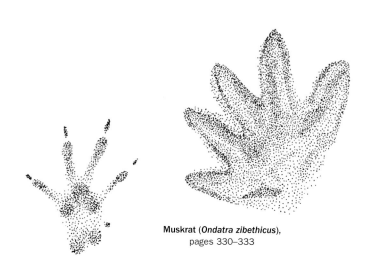

Muskrat (*Ondatra zibethicus*),
pages 330–333

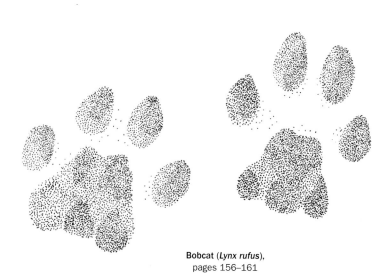

Bobcat (*Lynx rufus*),
pages 156–161

PLATE 22 Mammal Tracks: Actual Size

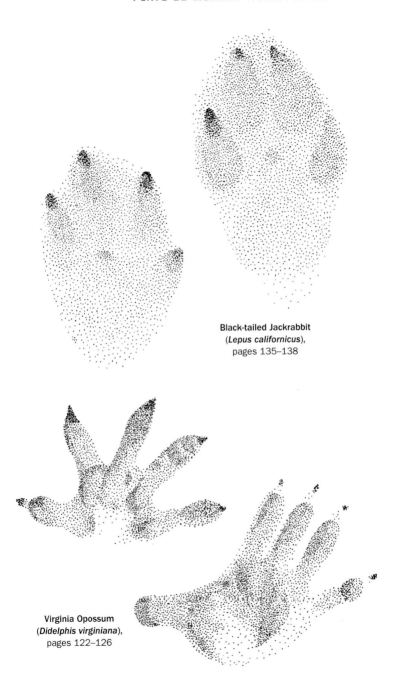

Black-tailed Jackrabbit
(*Lepus californicus*),
pages 135–138

Virginia Opossum
(*Didelphis virginiana*),
pages 122–126

PLATE 23 Mammal Tracks: Actual Size

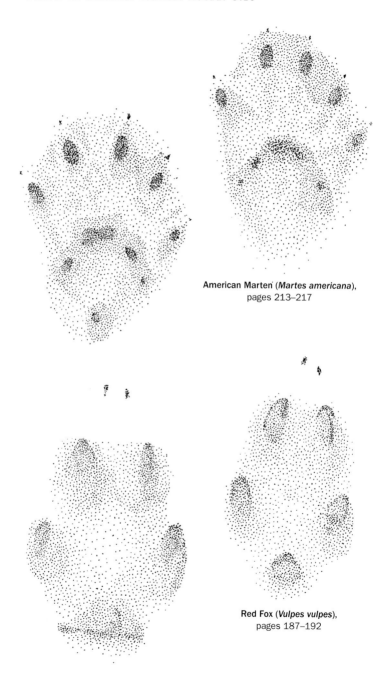

American Marten (*Martes americana*),
pages 213–217

Red Fox (*Vulpes vulpes*),
pages 187–192

PLATE 24 Mammal Tracks: Actual Size

American Badger (*Taxidea taxus*),
pages 231–234

PLATE 25 Mammal Tracks: Actual Size

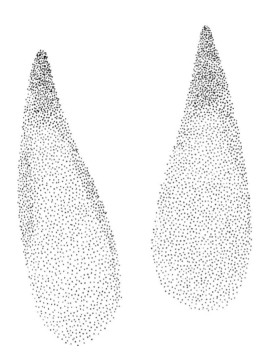

Mule Deer
(*Odocoileus hemionus*),
pages 257–260

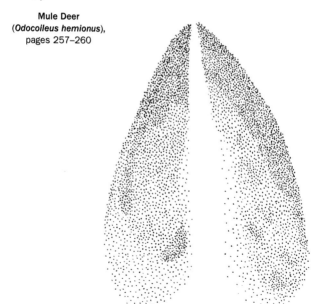

PLATE 26 Mammal Tracks: Actual Size

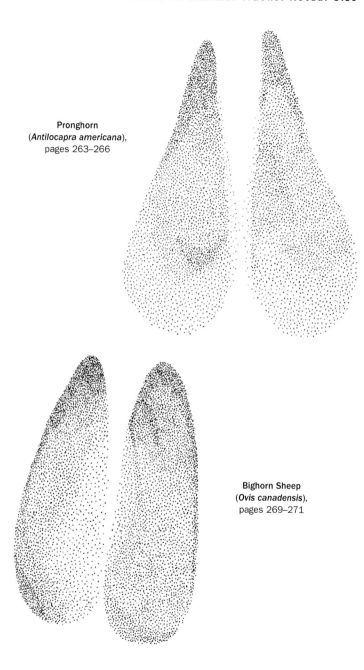

Pronghorn
(*Antilocapra americana*),
pages 263–266

Bighorn Sheep
(*Ovis canadensis*),
pages 269–271

PLATE 29 Mammal Tracks 50% of Actual Size

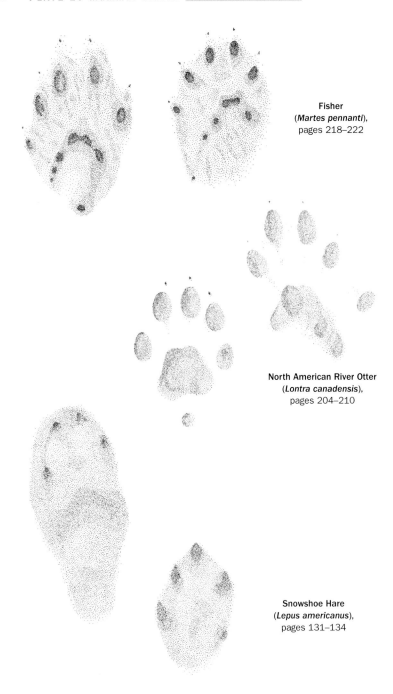

Fisher
(*Martes pennanti*),
pages 218–222

North American River Otter
(*Lontra canadensis*),
pages 204–210

Snowshoe Hare
(*Lepus americanus*),
pages 131–134

PLATE 30 Mammal Tracks 50% of Actual Size

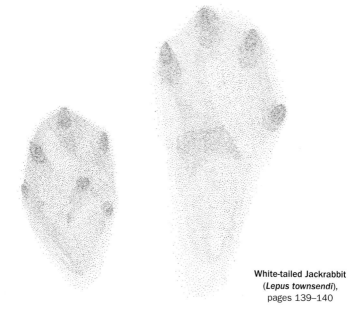

White-tailed Jackrabbit
(*Lepus townsendi*),
pages 139–140

Cougar, or Mountain Lion
(*Puma concolor*),
pages 162–167

PLATE 31 Mammal Tracks 50% of Actual Size

Wolverine (*Gulo gulo*),
pages 210–213

Coypu, or Nutria
(*Myocastor coypus*),
pages 355–358

PLATE 32 Mammal Tracks 50% of Actual Size

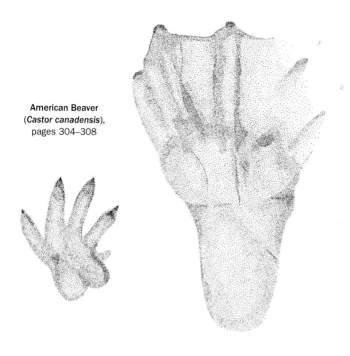

American Beaver
(*Castor canadensis*),
pages 304–308

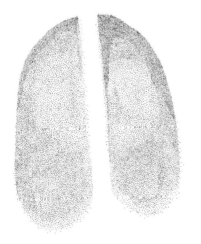

Elk (*Cervus elaphus*),
pages 260–263

PLATE 33 Mammal Tracks 50% of Actual Size

Wild Horse (*Equus caballus*),
pages 251–252

Wolf (*Canis lupus*),
pages 176–179

PLATE 34 Mammal Tracks 50% of Actual Size

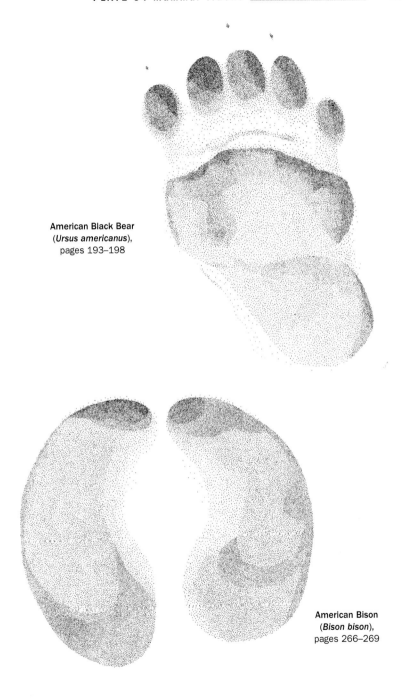

American Black Bear
(*Ursus americanus*),
pages 193–198

American Bison
(*Bison bison*),
pages 266–269

PLATE 35 Mammal Tracks 50% of Actual Size

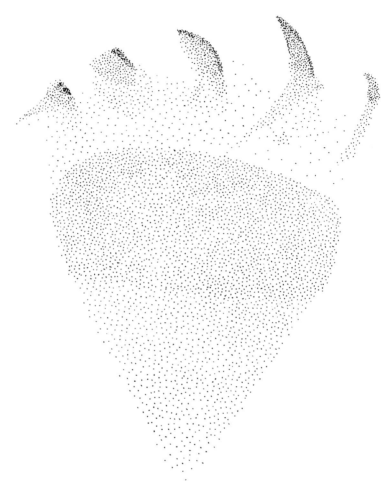

Harbor Seal (*Phoca vitulina*),
pages 199–200

PLATE 36 Bird Tracks: Actual Size

Classic Bird Tracks

Barn Swallow
(*Hirundo rustica*)
Length: 11/16–1 in.
(1.7–2.5 cm)
Hops

Redpoll
(*Acanthis* spp.)
Length: 7/8–1 in.
(2.2–2.5 cm)
Hops

Black-capped Chickadee
(*Poecile atricapillus*)
Length: 15/16–1 1/8 in.
(2.4–2.8 cm)
Hops

Common Yellowthroat
(*Geothlypis trichas*)
Length: 1–1 1/8 in.
(2.4–2.8 cm)
Hops

American Goldfinch
(*Spinus tristis*)
Length: 1–1 1/8 in.
(2.4–2.8 cm)
Hops

Belted Kingfisher
(*Megaceryle alcyon*)
Length: 1 1/8–1 1/4 in.
(2.8–3.1 cm)
Hops

Brewer's Sparrow
(*Spizella breweri*)
Length: 1 1/8–1 1/4 in.
(2.8–3.1 cm)
Hops

Winter Wren
(*Troglodytes troglodytes*)
Length: 1 1/8–1 1/4 in.
(2.8–3.1 cm)
Hops

PLATE 37 Bird Tracks: Actual Size

Common Ground-Dove
(*Columbina passerina*)
Length: 1 ⅛–1 ¼ in.
(2.8–3.1 cm)
Walks

House Sparrow
(*Passer domesticus*)
Length: 1 ⅛–1 ⁵⁄₁₆ in.
(2.8–3.3 cm)
Hops

Dark-eyed Junco (*Junco hyemalis*)
Length: 1 ¼–1 ½ in.
(3.2–3.8 cm)
Hops

White-throated Sparrow
(*Zonotrichia albicollis*)
Length: 1 ⁵⁄₁₆–1 ½ in.
(3.3–3.8 cm)
Hops

American Pipit (*Anthus rubescens*)
Length: 1 ⅜–1 ½ in.
(3.5–3.8 cm)
Walks

Horned Lark (*Eremophila alpestris*)
1 ⅜–1 ⅝ in. (3.5–4.1 cm)
Runs, walks

PLATE 38 Bird Tracks: Actual Size

Savannah Sparrow
(*Passerculus sandwichensis*)
Length: 1 3/8–1 5/8 in.
(3.5–4.1 cm)
Hops

Song Sparrow (*Melospiza melodia*)
Length: 1 3/8–1 5/8 in.
(3.5–4.1 cm)
Hops, skips

Snow Bunting
(*Plectrophenax nivalis*)
Length: 1 3/8–1 5/8 in.
(3.5–4.1 cm)
Hops, skips

White-breasted Nuthatch
(*Sitta carolinensis*)
Length: 1 1/2–1 5/8 in.
(3.5–4.1 cm)
Hops

Green-tailed Towhee
(*Pipilo chlorurus*)
Length: 1 9/16–1 3/4 in.
(4–4.4 cm)
Hops

PLATE 39 Bird Tracks: Actual Size

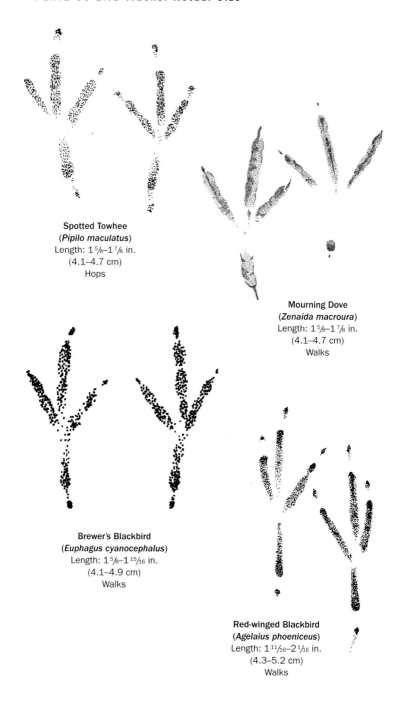

Spotted Towhee
(*Pipilo maculatus*)
Length: 1 5/8–1 7/8 in.
(4.1–4.7 cm)
Hops

Mourning Dove
(*Zenaida macroura*)
Length: 1 5/8–1 7/8 in.
(4.1–4.7 cm)
Walks

Brewer's Blackbird
(*Euphagus cyanocephalus*)
Length: 1 5/8–1 15/16 in.
(4.1–4.9 cm)
Walks

Red-winged Blackbird
(*Agelaius phoeniceus*)
Length: 1 11/16–2 1/16 in.
(4.3–5.2 cm)
Walks

PLATE 40 Bird Tracks: Actual Size

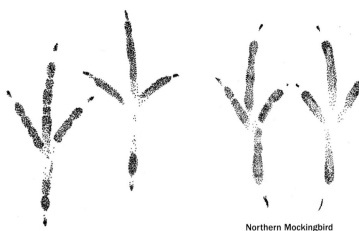

American Dipper
(*Cinclus mexicanus*)
Length: 1¾–1¹⁵⁄₁₆ in.
(4.4–4.9 cm)
Walks

Northern Mockingbird
(*Mimus polyglottos*)
Length: 1¾–2 in.
(4.4–5.1 cm)
Runs, hops

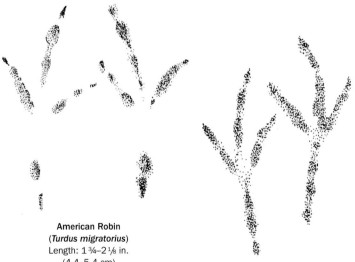

American Robin
(*Turdus migratorius*)
Length: 1¾–2⅛ in.
(4.4–5.4 cm)
Runs, skips

European Starling
(*Sturnus vulgaris*)
Length: 1¹³⁄₁₆–2⅛ in.
(4.6–5.4 cm)
Walks

PLATE 41 Bird Tracks: Actual Size

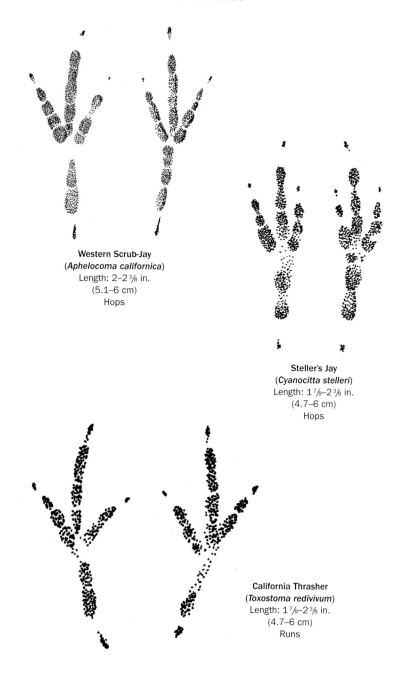

Western Scrub-Jay
(*Aphelocoma californica*)
Length: 2–2 3/8 in.
(5.1–6 cm)
Hops

Steller's Jay
(*Cyanocitta stelleri*)
Length: 1 7/8–2 3/8 in.
(4.7–6 cm)
Hops

California Thrasher
(*Toxostoma redivivum*)
Length: 1 7/8–2 3/8 in.
(4.7–6 cm)
Runs

PLATE 42 Bird Tracks: Actual Size

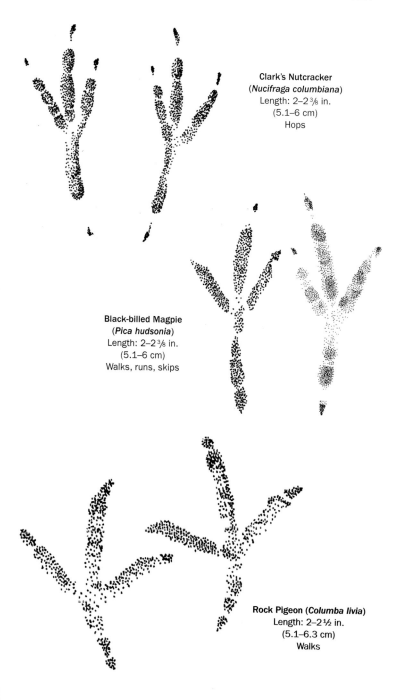

Clark's Nutcracker
(*Nucifraga columbiana*)
Length: 2–2 ⅜ in.
(5.1–6 cm)
Hops

Black-billed Magpie
(*Pica hudsonia*)
Length: 2–2 ⅜ in.
(5.1–6 cm)
Walks, runs, skips

Rock Pigeon (*Columba livia*)
Length: 2–2 ½ in.
(5.1–6.3 cm)
Walks

PLATE 43 Bird Tracks: Actual Size

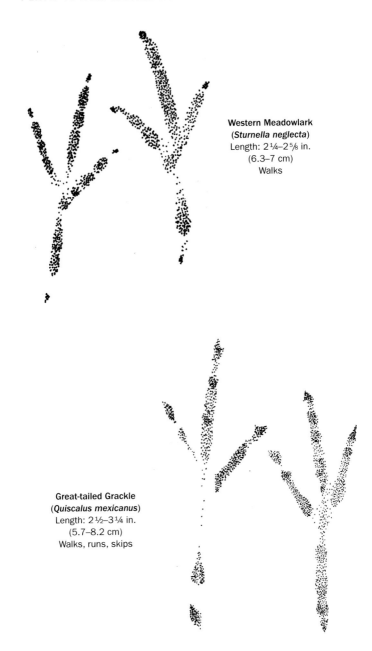

Western Meadowlark
(*Sturnella neglecta*)
Length: 2¼–2⅝ in.
(6.3–7 cm)
Walks

Great-tailed Grackle
(*Quiscalus mexicanus*)
Length: 2½–3¼ in.
(5.7–8.2 cm)
Walks, runs, skips

PLATE 44 Bird Tracks: Actual Size

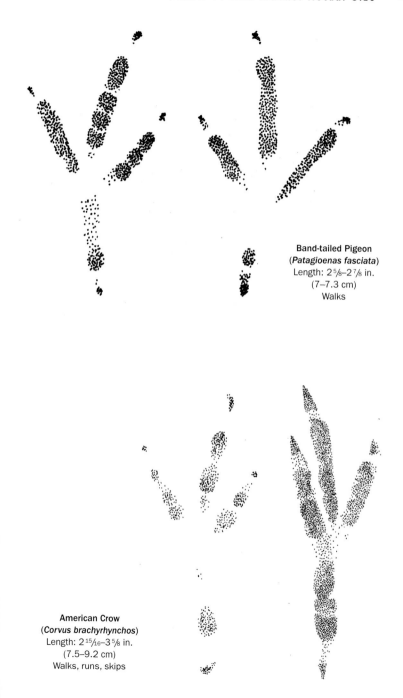

Band-tailed Pigeon
(*Patagioenas fasciata*)
Length: 2⅝–2⅞ in.
(7–7.3 cm)
Walks

American Crow
(*Corvus brachyrhynchos*)
Length: 2¹⁵⁄₁₆–3⅝ in.
(7.5–9.2 cm)
Walks, runs, skips

PLATE 45 Bird Tracks: Actual Size

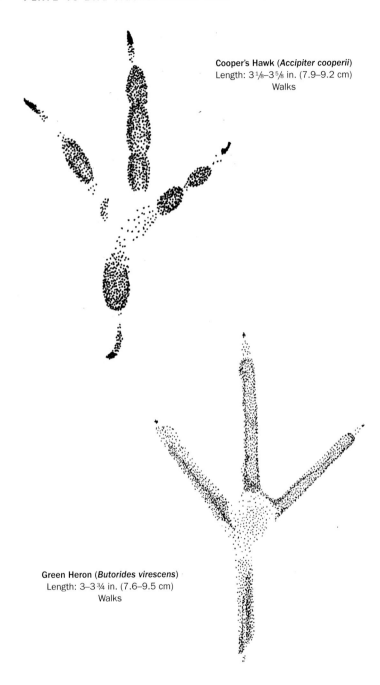

Cooper's Hawk (*Accipiter cooperii*)
Length: 3⅛–3⅝ in. (7.9–9.2 cm)
Walks

Green Heron (*Butorides virescens*)
Length: 3–3¾ in. (7.6–9.5 cm)
Walks

PLATE 46 Bird Tracks 50% of Actual Size

Common Raven
(*Corvus corax*)
Length: 3 ¾–5 in.
(9.5–12.7 cm)
Walks

Snowy Egret (*Egretta thula*)
Length: 3 ⅞–4 ⅝ in.
(9.8–11.7 cm)
Walks

Cattle Egret (*Bubulcus ibis*)
Length: 4–4 ⅝ in.
(10.2–11.7 cm)
Walks

Turkey Vulture
(*Cathartes aura*)
Length: 3 ¾–5 in.
(9.5–12.7 cm)
Walks, skips

PLATE 47 Bird Tracks 50% of Actual Size

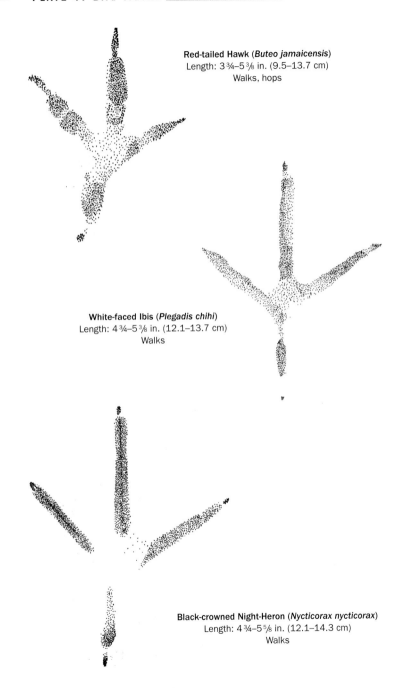

Red-tailed Hawk (*Buteo jamaicensis*)
Length: 3¾–5⅜ in. (9.5–13.7 cm)
Walks, hops

White-faced Ibis (*Plegadis chihi*)
Length: 4¾–5⅜ in. (12.1–13.7 cm)
Walks

Black-crowned Night-Heron (*Nycticorax nycticorax*)
Length: 4¾–5⅝ in. (12.1–14.3 cm)
Walks

PLATE 48 Bird Tracks 50% of Actual Size

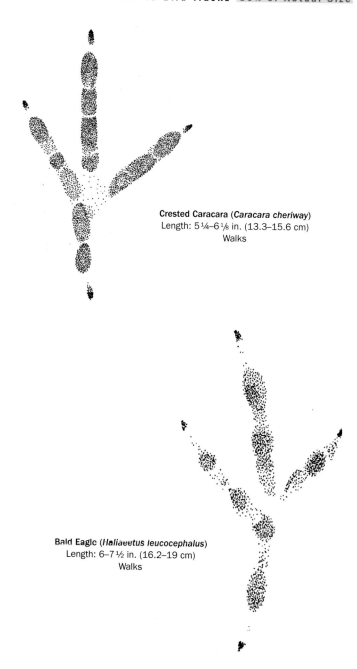

Crested Caracara (*Caracara cheriway*)
Length: 5¼–6⅛ in. (13.3–15.6 cm)
Walks

Bald Eagle (*Haliaeetus leucocephalus*)
Length: 6–7½ in. (16.2–19 cm)
Walks

PLATE 49 Bird Tracks 50% of Actual Size

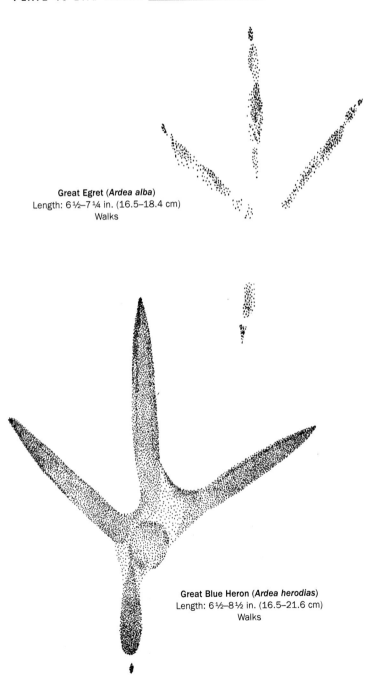

Great Egret (*Ardea alba*)
Length: 6½–7¼ in. (16.5–18.4 cm)
Walks

Great Blue Heron (*Ardea herodias*)
Length: 6½–8½ in. (16.5–21.6 cm)
Walks

PLATE 50 Bird Tracks: Actual Size

Game Bird Tracks

Common Poorwill
(*Phalaenoptilus nuttallii*)
Length: ¾–⅞ in.
(1.9–2.2 cm)
Walks

"Peep" sandpiper
(*Calidris* spp.)
Length: ¾–⅞ in.
(1.9–2.2 cm)
Walks, runs

Sanderling (*Calidris alba*)
Length: ¾–⅞ in.
(1.9–2.2 cm)
Runs, walks

Snowy Plover
(*Charadrius alexandrinus*)
Length: ¾–¹⁵⁄₁₆ in.
(1.9–2.2 cm)
Runs, walks

Spotted Sandpiper (*Actitis macularius*)
Length: ⅞–1 in. (2.2–2.5 cm)
Walks, runs

PLATE 51 Bird Tracks: Actual Size

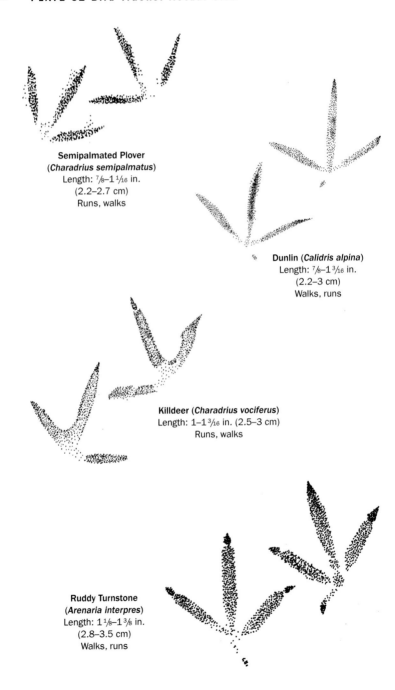

Semipalmated Plover
(*Charadrius semipalmatus*)
Length: 7/8–1 1/16 in.
(2.2–2.7 cm)
Runs, walks

Dunlin (*Calidris alpina*)
Length: 7/8–1 3/16 in.
(2.2–3 cm)
Walks, runs

Killdeer (*Charadrius vociferus*)
Length: 1–1 3/16 in. (2.5–3 cm)
Runs, walks

Ruddy Turnstone
(*Arenaria interpres*)
Length: 1 1/8–1 3/8 in.
(2.8–3.5 cm)
Walks, runs

PLATE 52 Bird Tracks: Actual Size

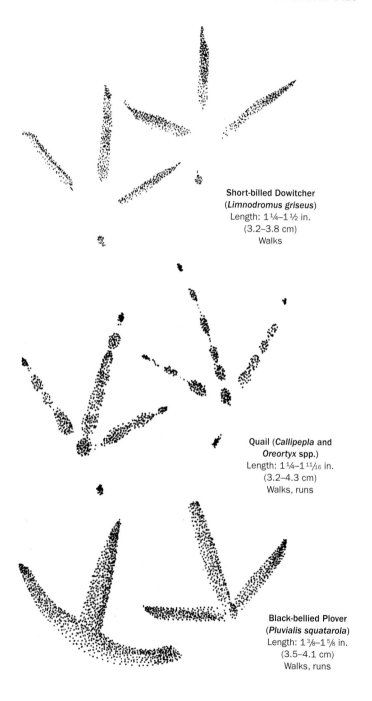

Short-billed Dowitcher
(*Limnodromus griseus*)
Length: 1 ¼–1 ½ in.
(3.2–3.8 cm)
Walks

Quail (*Callipepla* and
Oreortyx **spp.**)
Length: 1 ¼–1 ¹¹⁄₁₆ in.
(3.2–4.3 cm)
Walks, runs

Black-bellied Plover
(*Pluvialis squatarola*)
Length: 1 ³⁄₈–1 ⁵⁄₈ in.
(3.5–4.1 cm)
Walks, runs

PLATE 53 Bird Tracks: Actual Size

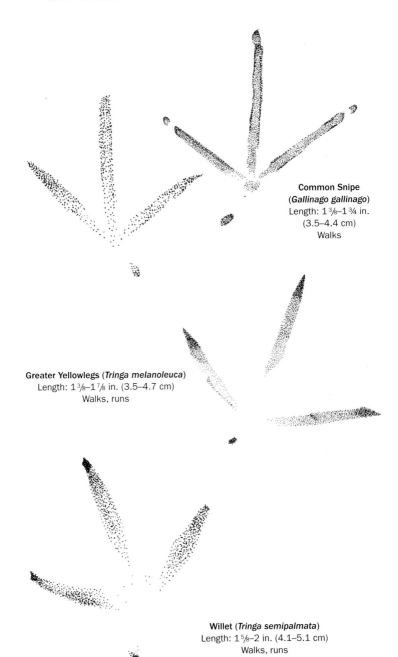

Common Snipe
(*Gallinago gallinago*)
Length: 1 3/8–1 3/4 in.
(3.5–4.4 cm)
Walks

Greater Yellowlegs (*Tringa melanoleuca*)
Length: 1 3/8–1 7/8 in. (3.5–4.7 cm)
Walks, runs

Willet (*Tringa semipalmata*)
Length: 1 5/8–2 in. (4.1–5.1 cm)
Walks, runs

PLATE 54 Bird Tracks: Actual Size

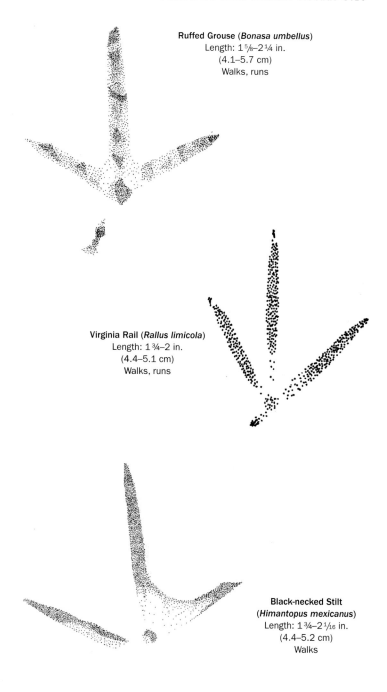

Ruffed Grouse (*Bonasa umbellus*)
Length: 1 ⅝–2 ¼ in.
(4.1–5.7 cm)
Walks, runs

Virginia Rail (*Rallus limicola*)
Length: 1 ¾–2 in.
(4.4–5.1 cm)
Walks, runs

Black-necked Stilt
(*Himantopus mexicanus*)
Length: 1 ¾–2 ¹⁄₁₆ in.
(4.4–5.2 cm)
Walks

PLATE 55 Bird Tracks: Actual Size

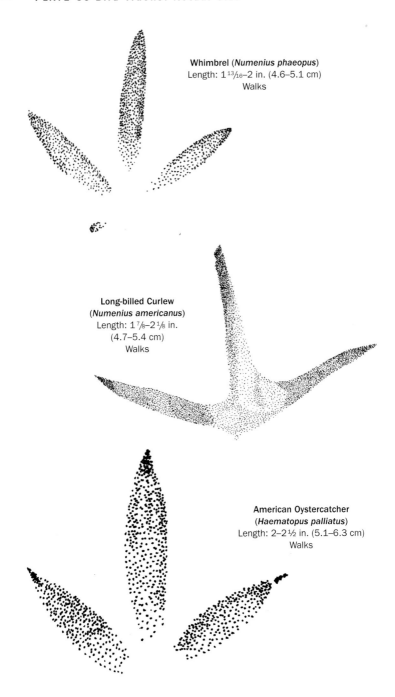

Whimbrel (*Numenius phaeopus*)
Length: 1 13/16–2 in. (4.6–5.1 cm)
Walks

Long-billed Curlew
(*Numenius americanus*)
Length: 1 7/8–2 1/8 in.
(4.7–5.4 cm)
Walks

American Oystercatcher
(*Haematopus palliatus*)
Length: 2–2 1/2 in. (5.1–6.3 cm)
Walks

PLATE 56 Bird Tracks: Actual Size

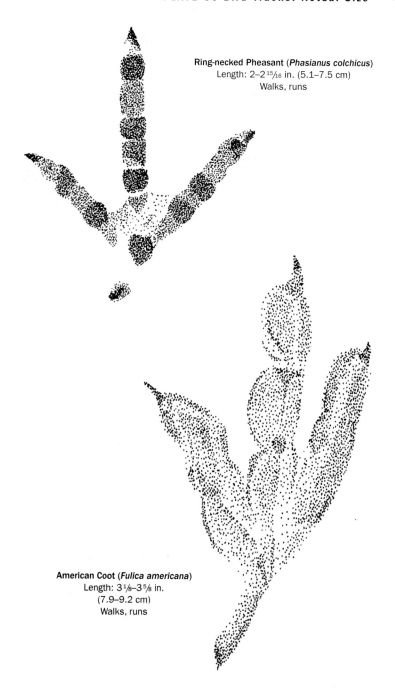

Ring-necked Pheasant (*Phasianus colchicus*)
Length: 2–2 15/16 in. (5.1–7.5 cm)
Walks, runs

American Coot (*Fulica americana*)
Length: 3 1/8–3 5/8 in.
(7.9–9.2 cm)
Walks, runs

PLATE 57 Bird Tracks 50% of Actual Size

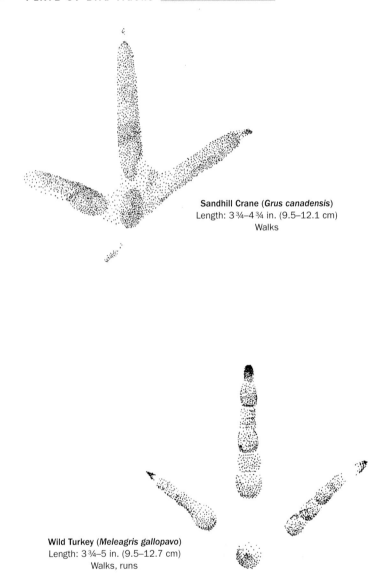

Sandhill Crane (*Grus canadensis*)
Length: 3 ¾–4 ¾ in. (9.5–12.1 cm)
Walks

Wild Turkey (*Meleagris gallopavo*)
Length: 3 ¾–5 in. (9.5–12.7 cm)
Walks, runs

PLATE 58 Bird Tracks: Actual Size

Palmate (Webbed) Tracks

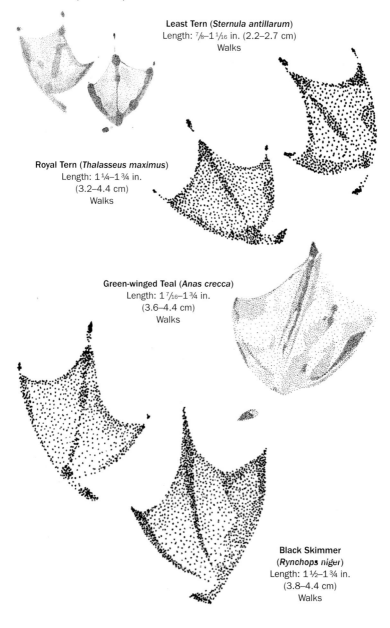

Least Tern (*Sternula antillarum***)**
Length: ⁷⁄₈–1 ¹⁄₁₆ in. (2.2–2.7 cm)
Walks

Royal Tern (*Thalasseus maximus***)**
Length: 1 ¼–1 ¾ in.
(3.2–4.4 cm)
Walks

Green-winged Teal (*Anas crecca***)**
Length: 1 ⁷⁄₁₆–1 ¾ in.
(3.6–4.4 cm)
Walks

**Black Skimmer
(***Rynchops niger***)**
Length: 1 ½–1 ¾ in.
(3.8–4.4 cm)
Walks

PLATE 59 Bird Tracks: Actual Size

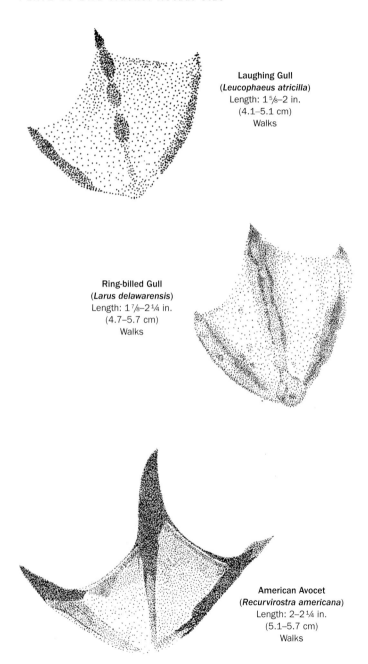

Laughing Gull
(*Leucophaeus atricilla*)
Length: 1 ⅝–2 in.
(4.1–5.1 cm)
Walks

Ring-billed Gull
(*Larus delawarensis*)
Length: 1 ⅞–2 ¼ in.
(4.7–5.7 cm)
Walks

American Avocet
(*Recurvirostra americana*)
Length: 2–2 ¼ in.
(5.1–5.7 cm)
Walks

PLATE 60 Bird Tracks: Actual Size

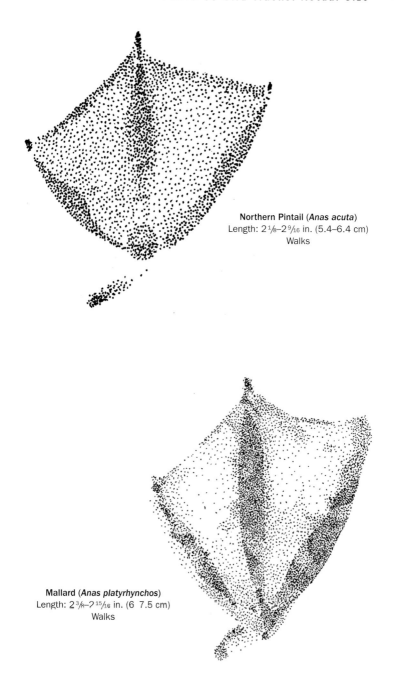

Northern Pintail (*Anas acuta*)
Length: 2 1/8–2 9/16 in. (5.4–6.4 cm)
Walks

Mallard (*Anas platyrhynchos*)
Length: 2 3/8–2 15/16 in. (6 7.5 cm)
Walks

PLATE 61 Bird Tracks: Actual Size

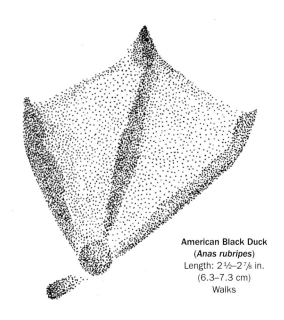

American Black Duck
(*Anas rubripes*)
Length: 2½–2⅞ in.
(6.3–7.3 cm)
Walks

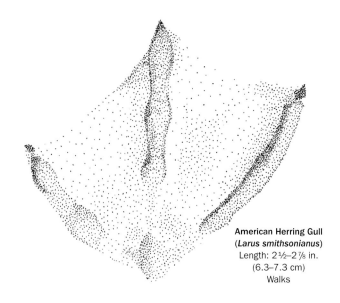

American Herring Gull
(*Larus smithsonianus*)
Length: 2½–2⅞ in.
(6.3–7.3 cm)
Walks

PLATE 62 Bird Tracks 50% of Actual Size

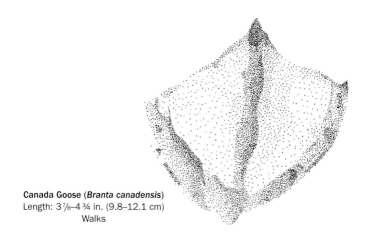

Canada Goose (*Branta canadensis*)
Length: 3 ⅞–4 ¾ in. (9.8–12.1 cm)
Walks

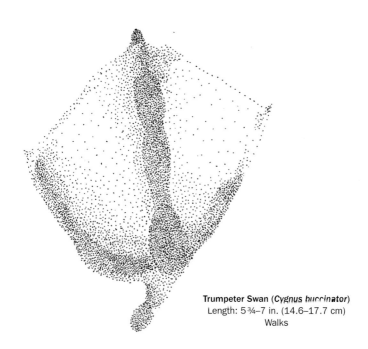

Trumpeter Swan (*Cygnus buccinator*)
Length: 5 ¾–7 in. (14.6–17.7 cm)
Walks

PLATE 63 Bird Tracks 50% of Actual Size

Totipalmate Tracks

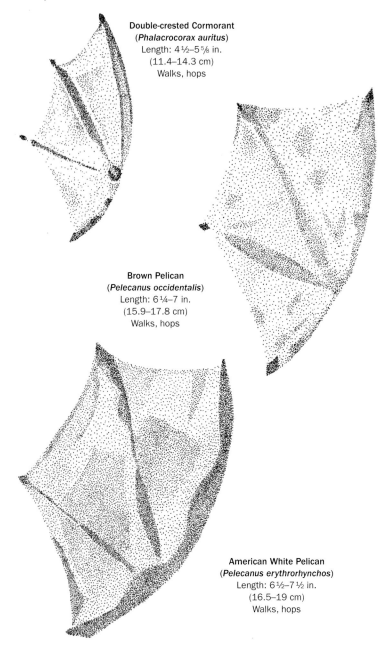

Double-crested Cormorant
(*Phalacrocorax auritus*)
Length: 4½–5⅝ in.
(11.4–14.3 cm)
Walks, hops

Brown Pelican
(*Pelecanus occidentalis*)
Length: 6¼–7 in.
(15.9–17.8 cm)
Walks, hops

American White Pelican
(*Pelecanus erythrorhynchos*)
Length: 6½–7½ in.
(16.5–19 cm)
Walks, hops

PLATE 64 Bird Tracks: Actual Size

Zygodactyl Tracks

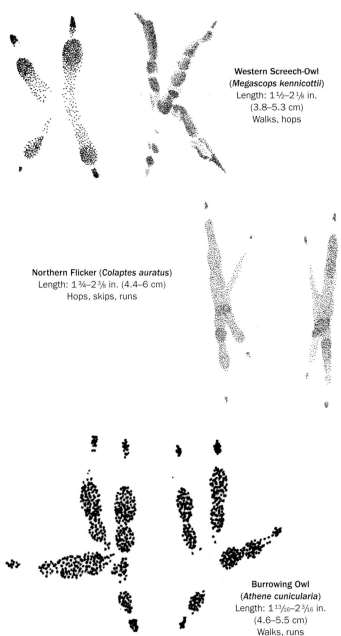

Western Screech-Owl
(*Megascops kennicottii*)
Length: 1½–2⅛ in.
(3.8–5.3 cm)
Walks, hops

Northern Flicker (*Colaptes auratus*)
Length: 1¾–2⅜ in. (4.4–6 cm)
Hops, skips, runs

Burrowing Owl
(*Athene cunicularia*)
Length: 1¹³⁄₁₆–2³⁄₁₆ in.
(4.6–5.5 cm)
Walks, runs

PLATE 65 Bird Tracks: Actual Size

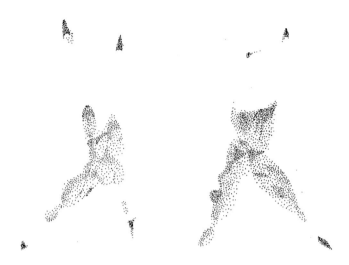

Pileated Woodpecker (*Dryocopus pileatus*)
Length: 2¼–2⅝ in. (5.7–6.6 cm)
Hops

Greater Roadrunner (*Geococcyx californianus*)
Length: 2⅝–3½ in. (6.7–8.9 cm)
Runs, walks, skips

PLATE 66 Bird Tracks 50% of Actual Size

Great Horned Owl (*Bubo virginianus*)
Length: 3¼–5 in. (8.2–12.7 cm)
Walks, hops

Snowy Owl (*Bubo scandiacus*)
Length: 3¼–5¼ in. (8.2–13.3 cm)
Walks, hops

PLATE 67 Reptile and Amphibian Trails: Actual Size

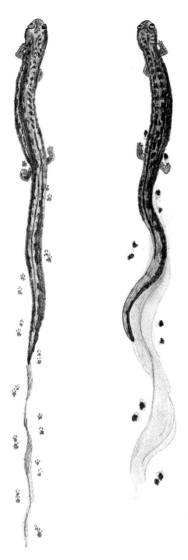

California Slender Salamander
(*Batrachoseps attenuatus*)
Trail width: Up to ⅝ in. (1.5 cm)

Yellow-blotched Ensatina
(*Ensatina eschscholtzii croceater*)
Trail width: Up to 1½ in. (3.4 cm)

PLATE 68 Reptile and Amphibian Trails: Actual Size

Horned lizard (*Phrynosoma* spp.)
Trail width: Up to 2¾ in. (7 cm)

PLATE 69 Reptile and Amphibian Trails: Actual Size

California Newt (*Taricha torosa*)
Trail width: Up to 2½ in. (6.4 cm)

PLATE 70 Reptile and Amphibian Trails 50% of Actual Size

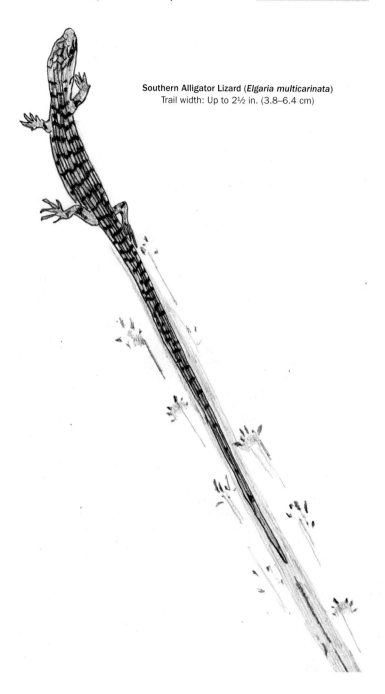

Southern Alligator Lizard (*Elgaria multicarinata*)
Trail width: Up to 2½ in. (3.8–6.4 cm)

PLATE 71 Reptile and Amphibian Trails 50% of Actual Size

Western Whiptail
(*Aspidoscelis tigris*)
Trail width: Up to 4 in. (10.2 cm)

California Giant Salamander
(*Dicamptodon ensatus*)
Trail width: Up to 3¼ in. (8.3 cm)

PLATE 72 Reptile and Amphibian Trails 50% of Actual Size

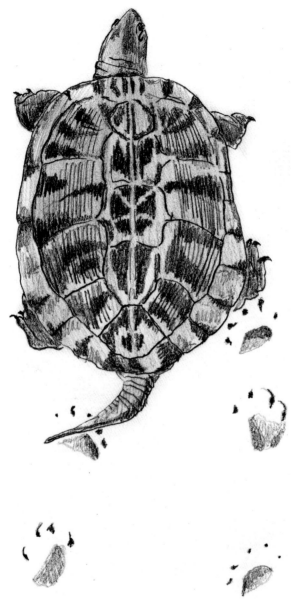

Western Pond Turtle, or Pacific Pond Turtle
(*Actinemys marmorata*)
Trail width: Up to 8 in. (20 cm)

PLATE 73 Reptile and Amphibian Trails 25% of Actual Size

Sidewinder
(*Crotalus cerastes*)
Trail width: 8–16 in.
(20–41 cm)

PLATE 74 Invertebrate Trails: Actual Size

Pill bug, or sow bug
(family Armadillidiidae)

Ironclad beetle
(*Nosoderma* spp.)

PLATE 75 Invertebrate Trails: Actual Size

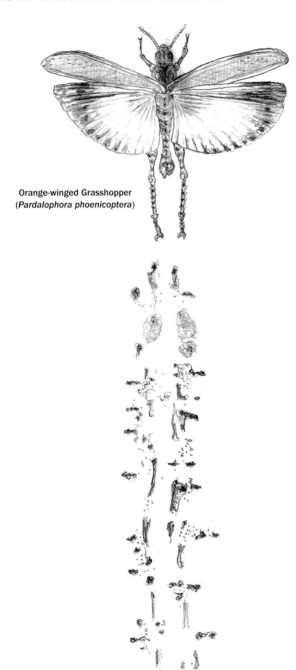

Orange-winged Grasshopper
(*Pardalophora phoenicoptera*)

PLATE 76 Invertebrate Trails: Actual Size

Calosomas beetle
(*Calosoma* spp.)

Black Burying Beetle, or Carrion Beetle
(*Nicrophorus nigritus*)

PLATE 77 Invertebrate Trails: Actual Size

Large darkling beetle
(*Coelocnemis* spp.)

PLATE 78 Invertebrate Trails 75% of Actual Size

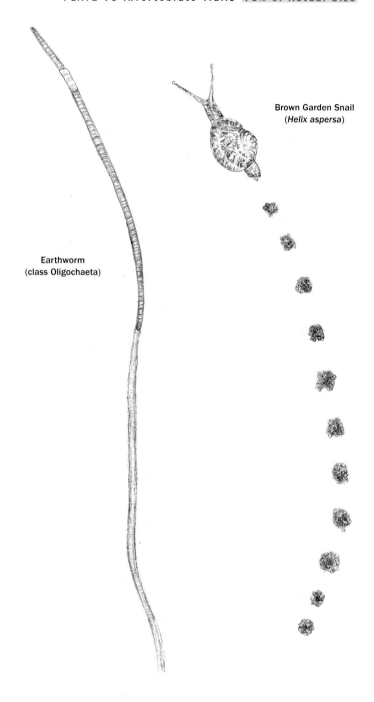

Brown Garden Snail
(*Helix aspersa*)

Earthworm
(class Oligochaeta)

PLATE 79 Invertebrate Trails 50% of Actual Size

Painted Tiger Moth larva
(*Arachnis picta*)

Millipede
(*Hiltonius* spp.)

PLATE 80 Invertebrate Trails   50% of Actual Size

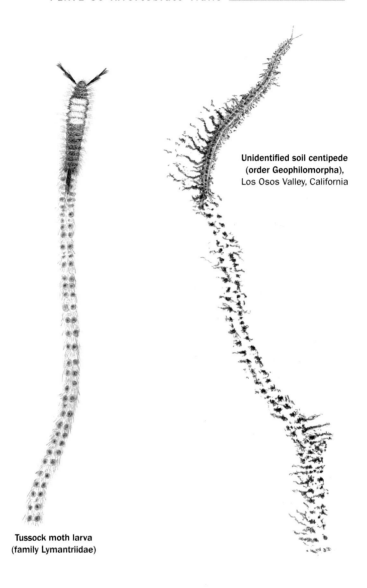

Unidentified soil centipede
(order Geophilomorpha),
Los Osos Valley, California

Tussock moth larva
(family Lymantriidae)

PLATE 81 Invertebrate Trails 50% of Actual Size

Jerusalem cricket, or potato bug
(*Stenopelmatus* spp.)

PLATE 82 Invertebrate Trails 50% of Actual Size

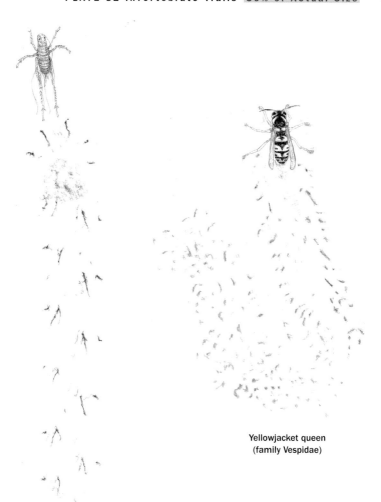

Yellowjacket queen
(family Vespidae)

California Camel Cricket
(*Ceuthophilus californianus*)

PLATE 83 Invertebrate Trails 50% of Actual Size

Wolf spiders (family Lycosidae) and
grass spiders (family Agelenidae)
Trail width: Up to 2½ in. (6.4 cm),
sometimes more

Mojave Desert Scorpion
(*Paruroctonus arenicola*)

PLATE 84 Invertebrate Trails 50% of Actual Size

California tarantula
(*Aphonopelma* spp.)
Trail width: 4–5½ in. (10.2–14 cm)

ANIMAL SCATS, URINE, AND OTHER
SCENT-MARKING BEHAVIORS

Mark Elbroch, Jonah Evans

Animal scat, urine, and other secretions offer a wealth of information beyond proving the presence of a species. They often constitute chemical communications and animal behaviors, as well as providing insights into animal diets and natural history. Overcome any preconceived queasiness you might have and move in for a closer inspection. Terminology for feces, excrement, pellets, or droppings is colorfully varied, but *scat* is the general term used by biologists and trackers to describe the solid waste materials excreted by animals in general, and mammals in particular. The term *droppings* is generally used when referring to bird scat, and the term *frass* is a general descriptor for insect scat, as well as a specific term for caterpillar feces.

Uric acid is toxic waste as well, and all animals must remove it from their bodies. Mammals and some reptiles and amphibians water it down and void it as liquid waste called *urine*. Birds and some reptiles employ a radically different approach, and instead concentrate uric acid and then dispose of it with their scats.

Many mammals have glands that can secrete chemicals directly into their urine, and then use urine scent posts in intra- and interspecies communication. Animals also employ numerous other glandular secretions and behaviors in chemical communication, some of which are visible to the human eye, others detectable to the nose, and others so subtle that only interpreting the trail will allow you to know that an animal has scent marked.

This chapter is predominantly a discussion of animal scats, rather than urine or other secretions. This is partly a result of our own limitations, the difficulty in finding such signs, and the difficulty in studying these behaviors in wild animals. Urine and other secretions are rich in information, but they are much subtler and have not received the attention from trackers and researchers they justly deserve.

Scat

Animal scat is filled with a fascinating array of indigestible substances, including hair, feathers, bones, insect limbs and carapaces, fruit skins, and other plant remains. Be hygienic and practice safety to protect yourself, but do not be afraid to break open and explore scat contents. Opening scats provides real information about a species' dietary habits, useful clues in species identification, and evidence of ecological relationships between the animal and its environment—such as predator-prey relationships or where the animal has foraged on the landscape. For example, you may find a meadow with abundant vole sign and scat. If you look closely at these scats, you will find that they are filled with vegetation. Upon further exploration, you might discover fox and Coyote scats, and maybe barn owl pellets filled with vole skulls and hair in the adjacent forest. This short exploration reveals the ecological importance of the meadow—not only to voles, but also to a variety of other species through the food chain. Signs of predator and prey are identified together.

SAFETY CONSIDERATIONS WHEN HANDLING SCAT

There are a variety of dangerous pathogens and parasites in wildlife feces that are transmissible to humans, and that can pose significant health risks or even death. Although the chance of contracting a disease from tracking wildlife and studying their feces is *highly* unlikely, we strongly advise trackers to practice a few basic safety precautions when handling scats:

Always wear gloves or thoroughly wash your hands after handling scat or animal remains. One of the primary disease and parasite transmission modes is through ingestion of microscopic quantities of feces from an infected animal and/or contact with blood.

The other transmission mode that concerns trackers is inhalation of contaminated dust particles containing fecal matter. Old, dried scats are more dangerous than fresh ones, because they are more likely to create dust and some pathogens can remain viable for many years.

The following is a sample of some of the more common zoonotic (transmissible from animals to humans) diseases and parasites. We gathered this information from (and highly recommend) the books *Infectious Diseases of Wild Mammals* by Williams and Barker and *Parasitic Diseases of Wild Mammals* by Samuel *et al.* Also, the Centers for Disease Control and Prevention maintain a detailed website with information on all the following diseases (http://www.cdc.gov).

Hantavirus and Arenavirus. These two functionally similar viruses are carried primarily by mice and rats and transmitted to humans through contact with their scat or inhalation of contaminated dust. Hantavirus infections cause a respiratory disease called hantavirus pulmonary syndrome that produces flulike symptoms in humans, including fever, muscle aches, dizziness, paralysis, convulsions, coma, and death. *Avoidance:* Infection occurs when you breathe in virus particles, so it is important to avoid actions that raise dust, such as sweeping or vacuuming, where rodents have been active. When you need to clean out an old shed or cabin littered with mice droppings, *first* spray all surfaces with a bleach-water mix, to kill the virus. Wait until the surface is dry, and then sweep and clean as normal.

Salmonellosis. The *Salmonella* bacterium is carried in numerous mammals, livestock, reptiles, and birds, and is transmitted to humans through ingestion of fecal matter or contaminated water or food. Symptoms include fever, nausea, diarrhea, and vomiting.

Tularemia. The causative bacterium (*Pasteurella tularensis*) is carried primarily in rodents, rabbits, and hares, and is transmitted to humans through numerous means, including the bites of flies and the handling of carcasses and feces. Effects run the gamut from mild illness to death. Note: this disease is extremely rare in the United States.

Giardiasis. The protozoan *Giardia* is carried in a wide variety of species and transmitted to humans through ingestion of fecal matter or contaminated food or water. It is also highly contagious and is easily contracted from an infected person. Symptoms include diarrhea, stomach cramps, and nausea. *Avoidance:* Treat water (boiling, filters, or iodine tablets) before drinking.

Histoplasmosis. The causative fungus can be found in soil areas contaminated by bat and bird droppings. It is transmitted to humans through inhalation of the spores when contaminated soil is disturbed. Symptoms may be nonexistent or may include fever, chest pain, dry cough, and death if untreated. *Avoidance:* When investigating bat roosts, wear masks and other protective clothing.

Toxocariasis. This infection is caused by two species of *Toxocara*, parasitic roundworms primarily carried in domestic dogs and cats. It is transmitted to humans primarily through ingestion of eggs found in feces or eating undercooked meat. Symptoms include coughing, fever, asthma, pneumonia, and blindness. *Avoidance:* Use gloves to handle pet scats, and do not eat pets unless they are well cooked.

Baylisascaris. This is a genus of roundworms primarily known to infect raccoons but also found in more than 90 other species including dogs, foxes, bears, skunks, badgers, fishers, martens, rodents, and birds. It is transmitted to humans through ingestion of eggs found in feces. Symptoms include skin irritation, nausea, eye damage, severe brain damage from random migrations of the larvae through the brain, and death. *Avoidance:* Handle raccoon scats with gloves and do not smell them, especially when they are completely dried and spores may be displaced and take to air more easily. Be careful of latrine sites, which raccoons sometimes make in children's sandboxes; children playing can accidently ingest eggs in the sand.

The colorful fresh scat of a Coyote eating fruits and greens.

Limitations in Interpreting Animal Scats

Animal scats vary tremendously depending upon seasonal food availability, contents, age, and the health of the animal, and are thus sometimes challenging to identify in the field and difficult to describe in categories.

However, expertise can be achieved with practice. Zuercher, Gibson, and Stewart (2003) used genetic techniques to test the ability of indigenous hunters in Paraguay to identify the various mammal scats on local trails to determine whether local people and indirect methods could be used to document the local mammalian assemblage. Indigenous hunters proved 100 percent accurate in scat identification. Thus, practice and experience will help you identify most scats you encounter, but come to peace with the fact that there will forever be scats that cannot be positively identified because of visual variations from diet, health, or other stresses on the animal. It is wiser to narrow down a scat to a few possible species than to misclassify it. Newer genetic techniques to identify species from their scats continue to improve and drop in price, allowing for a useful check to a tracker's ability to identify scats correctly.

Laboratory analysis of scat contents has also been found to have limitations and thus can bias what we believe we know about animal foraging and diets. Work by Gamberg and Atkinson (1988) showed that not everything consumed by ermines was reliably detectable in ermine scats. Lynn Rogers (personal communication) showed that wetland plants pass through black bears without being detected. Researchers such as Weaver (1993), who studied Gray Wolf scats, have created equations that interpret a scat's contents to better estimate what an animal actually ate and how much of each food they ate. Such equations have been researched and published for numerous species and are a useful and necessary calibration when you attempt to interpret what and how much an animal eats from food remains passed in scats. Rapid advancements in genetic technologies, such as DNA bar coding, are also expanding our ability to identify food remains, and we may someday soon be able to send a scat to a lab and know all of its contents with certainty.

Chemical Communication

Scats, urines, and other scent-marking behaviors are part of an intricate communication system, potentially relaying information not only about the age, health, sex, and reproductive receptiveness of the individual, but also to mark territorial boundaries, to establish ownership, or to contest the presence of another animal. We do not completely understand what scats and urine mean when posted and placed with great care by wild animals. Hope Ryden (1989) offers us a glimpse of what all these markers may mean in her *Lily Pond: Four Years with a Family of Beavers.* She observed a resident male's response to a trespassing beaver's scent mound. The male mounted the shore hissing and growling. He destroyed the scent mound with his front paws and carried a portion of it to the water's edge, sinking it in the pond. Other researchers have recorded similar responses in beavers (Wilsson 1971; Müller-Schwarze and Heckman 1980), as well as observed residents building their own scent mounds atop a trespasser's, or secreting their own castoreum in an attempt to cover the scent of a passing beaver.

A pair of beaver scent mounds along a pond edge.

We do know that the olfactory sense of wild creatures can be many hundreds of times more sensitive than our own; what we depend upon our eyes to decipher and interpret in the environment, many mammals can achieve with eyes shut and nostrils flaring.

When Predators Eat Large Prey

When a predator takes large prey, one that requires several feedings to finish, scat contents hint as to where on the carcass the animal was recently feeding. Predators tend to eat the internal organs first, which result in black, moist, soft scats with little bone or fur. As they continue to feed, the scats incorporate more bone and fur, and in the end, will be composed of these items entirely. For example, when tracking Cougars to their kills, you will find that the latrine adjacent to the kill or nearby en route to or next to their bedding site, will contain loose, black scats composed of the internal organs and meat. It is only when you continue to follow the cat after it has abandoned the carcass completely, as it fans out to circle and patrol its territory, that you find scats of bone and fur on some regular travel route it uses. Then it locates and stalks another large prey, and the cycle begins anew.

Scats also hint at status in Wolf and Coyote packs. The choicest parts of the kill, the internal organs, are eaten by the alpha pair. Depending upon pack status, the quality and quantity of good meat drops substantially, and those at the bottom of the pecking order may feed much more on hide and bones than those at the top (Peterson and Ciucci 2003).

A Cougar scat composed entirely of meat and internal organs, turned completely white and crumbling after a week.

Physical Characteristics of Scats and Droppings Useful in Identification

Relative Size

The size of scats is inherently misleading, given the tremendous individual and within-species variation due to diet. Certainly size is very useful in differentiating between the scats of bears and mice, but it is less useful when differentiating between two species of similar size. There is substantial overlap in the measurements of scats across species. This is especially true when comparing scat lengths; scat diameters (the widths) are more reliably diagnostic, because this dimension is more predetermined by the internal excretory structures of the animal responsible. Even diameter measurements, however, provide only a coarse clue to the identity of a species.

With these warnings well in place, scat measurements can be useful, as can be a rough approximation of volume—meaning how much scat has been deposited. Judging the volume of a scat becomes easier with experience. Once you've identified fifty Bobcat scats, you'll probably identify your first mature Cougar scat with ease. Measurements should always be used in combination with other characteristics: shape, color, texture, contents, odor, and location at varying scales—meaning within the immediate vicinity of the scat, such as atop a rock, as well as within the larger habitat. Within their natural setting, surrounded in contextual clues, scat identification becomes much more manageable, and fun.

Uric Acid

Is there uric acid present—visible as either a completely white packet separate from the scat, a wash of white coating the scat, or even a streak or splat of white liquid?

Uric acid can be critical to positively identifying certain species. Some useful questions to further characterize the uric acid fall in the next section starting on page 108.

A Red Fox scat composed of apple skins and pulp placed carefully atop a rock in a hiking trail.

Color

Dietary variation and age will influence the color of scats. As described above, carnivore scats vary depending upon what part of a carcass they are feeding. Fruits dye scats a rainbow of colors, and a variety of other contents retain their color through digestion.

Study the color of both the exposed portion of scat and the hidden portion found beneath, as well as the color of the ground or vegetation that is covered by the scat. Scats generally dry and harden with age, whitening as they are left in the open for long periods. (Scats may also be white due to diet.) At Point Reyes National Seashore, a very fresh scat of a Bobcat was moist, large, black, and dark green. In just three hours of sun exposure, the outer portions had completely dried and the entire scat had become a very light tan color; the inside still held moisture.

The area of ground or vegetation under the scat can darken and brown in less than five days, depending on other variables. Environmental conditions vary tremendously across California and are crucial factors when considering color and other aging factors.

Basic Shapes

Determine into which of the following three categories your scat falls: (1) tubes and cylinders, (2) pellets and tiny scats, or (3) blobs and squirts.

Tubes and cylinders: Tubes and cylinders can be one long, complete piece, or broken, segmented, and/or folded into many parts. These scats are characteristic of fruit and meat eaters.

Pellets: Lagomorphs (rabbits) leave pellets throughout the year. Deposited one at a time, a collection of rabbit pellets is an indication of the duration of time it spent in a location. Lizards and toads leave scats that resemble elongated pellets. Ungulates, including deer, leave pellets when browsing, and deposit a group of pellets at one time.

Blobs and squirts: Sometimes, meat or high-water-content fruit scats are amorphous and soft. Numerous game birds, like grouse, defecate two sorts of scats. Those produced in the lower gut are the more familiar cylindrical, pelletlike scats, while those produced in the upper gut—called the cecum—are amorphous, brown squirts easily confused with mammal scats. Often you find these scats in different roosts, but on occasion you will find them together.

More on Shape

TWISTS. Note whether a scat twists or remains straight along its length. Also note the degree to which the scat twists, as both mustelid (weasel) and canid (dog) scats twist when their scats are composed of fur and bone; in general, weasel scats are much more twisted and narrow. Cats, bears, and raccoons are among those whose scats generally do not appear twisted but instead have smooth surfaces.

TAPERED ENDS. Are the ends of the scat tapered? The most tapered end of the scat is the last to exit the anus of the animal. Canid scats often have one or two tapered ends. Weasel scats generally have very long tapered ends. Scats composed of hair are more tapered than those of fruits. Cat scats also occasionally have tapered ends, but when this happens it is often only one tapered end, the last to leave the animal. Blunt ends are more characteristic of the scats of felids, raccoons, wild boars, and bears.

SEGMENTS. Segments can vary from subtle narrower rings around the circumference of the scat to deep grooves that may cause the scat to break into completely separate sections. Cats often leave segmented scats, either joined or separated. Canids and mustelids tend to be one cord, though bent back on itself multiple times. Bear and raccoon scats break easily, leaving blunt ends and giving the impression of segmented scats, and bear scats composed of grasses and roots are often highly segmented.

Scat Contents

What's in the scat? Is it fruit skins, seeds, other vegetation, fur, bones, or a mystery mash? The contents of scats will obviously help you determine what animal might be responsible—that is if you have done your homework and know something about the diets of California wildlife. If you recognize the skins and seeds of apples, for instance, you can rule out felids immediately. If you identify fish scales, you can really begin to narrow your list of possibilities, granted some unlikely animal didn't scavenge a fish carcass.

Scent

Species have unique odors associated with their scats. Smells are very difficult to describe but relatively easy to distinguish with experience. Thus, you should smell every scat, unless there is a suspected health risk (see the section, "Safety Considerations When Handling Scat," at the start of this chapter), to begin to increase your catalog of odors associated with scats.

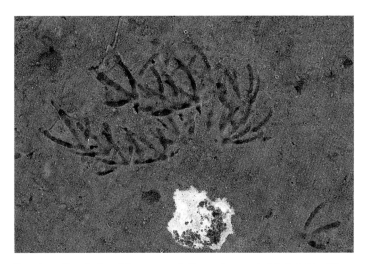

The uric portion of a Least Tern surrounds the small amount of dark, indigestible material defecated.

Most mammals have scent glands just inside the anus, and therefore scats serve the double purpose of digestive excretion and as a potential communicative device. Most birds in North America are not thought to have a sense of smell (a notable exception is the Turkey Vulture) and thus probably do not communicate much with scents. However, there are some intriguing recent studies that suggest the sense of smell in birds may be better developed than previously believed (Steiger, Fidler, and Valcu 2008).

Physical Characteristics of Uric Acid Useful in Identification

Distinct Packets

In some reptiles the uric waste is contained within a thin membrane, or "sac," that sits adjacent to the undigested remains in the scat. Lizards are the model species for this design; look for their two-part scats on regular perches on rocks or along fences. Geckos, too, leave similar scats clinging to walls where they inhabit houses, or rocks and trees where they live outdoors.

Like Paint Covering All or a Part of the Scat

In numerous herbivorous and insectivorous birds, and some reptiles including snakes, the uric acid is deposited as a white coating covering all or a part of the scat. In numerous birds, just one end of the scat looks as if it were dipped in white, whereas in other birds, such as sparrows and towhees, the white uric acid typically coats the entire scat. The amount of uric acid (meaning a thin layer or thick, gooey layer attached to the scat) can play an important role in scat identification as well. Roadrunner scats, for example, are easily identified by their large, thick uric acid cap at one end.

Streaks

Many raptors, meaning carnivorous birds, shoot uric acid from their cloacas rather than drop it straight down. They lean forward, raise their tail feathers, and project the uric acid as if from a squirt gun. The volume of uric acid is somewhat useful in species identification, but the lengths of the streaks are confounded by the height at which the bird was perched when it defecated. A Red-tailed Hawk perched atop a tree can produce a streak that measures 13 feet tip to tip, but when the same species of bird defecates while standing on the ground, the streaks measure only 3–3¾ feet (Elbroch and Marks 2001).

Cormorants also produce streaks rather than splats, making their defecations easy to pick out at a loafing area where they are intermixed with gulls and pelicans.

Splats

Other carnivorous and scavenging birds drop their defecations straight down, where they splat and puddle on the ground. Look carefully for small amounts of darker material as well, that bit of digested solid waste that was not formed into pellets and regurgitated via the mouth (see the chapter on bird pellets).

Owls, gulls, pelicans, and herons are among those species that make splats rather than streaks. Identifying splats versus streaks at kill sites where an avian predator has fed upon prey is especially useful in narrowing down the potential list of predators. Hawks and eagles make streaks, while owls drop their scats straight down.

Contextual Clues of Scats and Droppings Useful in Identification

Expanding our perspective, here are some useful clues in analyzing the larger area within which you have found your scats.

Scrapes

Bobcat and Cougar scrapes (see page 110), in which scats are sometimes placed or urine sprayed on the piled debris, can be quite elaborate and deep, or simply a swipe of a paw. Otters occasionally dig scrapes and drop their scat either in or next to it, and Golden-mantled Ground Squirrels also create small scrapes for both urine and scat deposits. Canids scratch near scats and scent posts from time to time, most often with the back feet. Pronghorn bucks also make territorial scrapes. Scrapes unassociated with scats or urine, but used in scent marking, are discussed within the species accounts at the end of the text.

The scratch marks that accompany the scat of a Coyote on hard ground.

Covered Scats

Felids do not always cover their scats, but from time to time they take great care in piling earth, snow, or available debris over their scats, and leave a ring of scratch marks about the pile. Little is known about why cats cover their scats sometimes and not others. Canid scat may also be partially covered from time to time. Coyotes that scratch after they defecate may unintentionally shower a scat behind them with debris.

Trails and Junctions

Canids are renowned for marking established runs, and their scats are common along hiking trails and roads. Felids also defecate in and along hiking trails, and on other paths worn by game species. Martens also mark hiking trails, making their scats much easier to find than the closely related Fisher, which more often defecates away from human traffic.

Raised Surfaces

Many species defecate on high points along travel routes. This is common for canids and mustelids on one scale, and for California Voles on a much smaller scale. Canids post scats on boulders, roots, mushrooms, bear scats, abandoned shoes, and an assortment of other elevated surfaces. Marten, Fisher, mink, and smaller weasels also place scats along raised surfaces. Fisher trails often veer off to a stump or rock, from several inches to several feet in height, only to pause long enough to deposit an elevated scat on their rounds. Martens do this too, and mink and muskrat scats are often found on raised vegetation clumps or rocks and logs in wetlands and streams where they are common.

Raccoons and Ringtails also leave scats raised above ground level. Both are excellent climbers, and scats are often found in tree crotches and along limbs where they have spent time. Be careful, because sometimes raccoons leave scats like land mines along limbs and in crotches, ideal spots for a hand when climbing trees and pulling oneself higher.

Raccoon scats lying atop a natural bridge over a small stream.

Natural Bridges and Funnels

"Bridges" are features that funnel animal movement. They can be fallen logs over water or tangled forest floor, including beaver dams bridging two land bodies or even rock walls and fences for some smaller species. Follow downed logs for a length of time, and you are bound to find scats of raccoons, Ringtails, martens, Fishers, weasels, Gray Foxes, or Douglas's Squirrels. Beaver dams should be regularly checked as well, because they are commonly marked by beavers, otters, Coyotes, Wolves, foxes, Bobcats, raccoons, and Fishers. Lizard, bird, and gray squirrel scats can often be found along fences. Even figurative land bridges, such as the narrowest points between two bodies of water, or a mountain pass between two valleys, are all places where animals move more frequently. For this reason they are a wonderful place to post scents, because these areas receive greater animal traffic and therefore offer the potential for increased attention.

Nests and Dens

Scats are also a clue that nests or dens are near or can help in interpreting nests and dens you find. Woodrat scats are always plentiful near burrows and nests, as are porcupine scats near their retreats. Fox kits almost always create a large latrine near their den entrances where tiny, twisted scats accumulate. Many birds carry chick droppings away from their nest a short distance and can sometimes lead the observant tracker to the nest.

When waters receded from this temporary pond adjacent to the Sacramento River in the dry season, a beaver latrine was exposed for view. The large tracks are a Great Blue Heron, and the smaller mammal, a Ringtail.

Scat Accumulations and Latrines

Trackers use the term *latrine* to describe a site where an accumulation of scat has been intentionally deposited by an individual or species over time. Scats can accumulate randomly, but a latrine *implies* that an animal intentionally chooses a particular spot over others to defecate and/or mark. Scat accumulations may simply indicate a greater length of time spent in one area, such as those found where jackrabbits forage, or where vulture, hawk, and owl droppings accumulate beneath nests and perches. Raccoons, bears, Coyotes, Bobcats, woodrats, voles, skunks, mustelids, muskrats, and beavers are among those that create intentional latrines.

Bird droppings also accumulate on the landscape, especially at loafing areas, regular perches, and nests. Accumulated white uric acid streaks and splats are called *white wash*, and can quite literally paint rocks, walls, trees, and ground. Cormorants, gulls, and others paint rocks and other regular perches where they rest and dry off. Nests of raptors are often ringed in white droppings, and below the nests of swifts and other rock-dwelling species, white wash can often be visible over long distances. The regular perches of owls, too, become covered in white, streaked with darker colors as well.

Aging series 1: Day 1, a fresh river otter scat.

Aging series 2: Day 4, in only three to four days of sun, an otter scat will crumble and whiten.

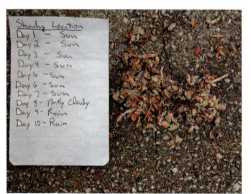

Aging series 3: Day 10, the color returns to the scat because of a brief rain shower—making aging based solely on oolor potentially problematic.

Aging Scats

As with tracks, determining the relative age of scats takes considerable practice and a good knowledge of local weather. Over time scats containing meat and organ tissue turn white, then crumble and eventually disappear. Scats containing hair and bones hold their form for longer periods of time—and the fur retains its color interminably, unless frequently exposed to sun. Fish scales and crayfish shell remains are visible long after a scat deposit, although the initial form of the scat deteriorates quickly.

Pellets formed of woody browse and bark persist many months, and are often seen well after winters when they were dropped. Research by Wigley and Johnson (1981) proved the importance of moisture in deer pellet disintegration; during the wet season in the southeastern United States deer pellets disappeared within a month, while pellets dropped in the drier conditions found in the West were identifiable well over a year after they were made. With practice and an understanding of the effects of water and wind on scats in the area where you track, aging scats can be identified with greater certainty. But be aware that a scat moved only a few feet from the sun to the shade can drastically influence its rate of decay.

Urine and Other Scent-Marking Behaviors

Like scat, urine and other secretions are used for complex communications between species and individuals, and are invaluable to the tracker and researcher. Mech, Seal, and Delgivdice (1987) showed the usefulness of collecting urine samples while winter-tracking wolves. Analysis of a single urine sample revealed how recently a wolf had eaten, and an analysis of several samples revealed the health of the entire pack!

The contextual clues surrounding urine and other scents are tremendously important. The very same contextual clues described previously for scat interpretation, apply also to urine and other secretions. Remember that you may have to use your sense of smell rather than your eyes, because urine is less visible than scat. When a Bobcat cuts sideways to walk close to a small rotting stump, your nose may tell you what your eyes cannot see.

Urine and some secretions have distinctive smells, and our olfactory memory can be extremely useful for species identification. If you've ever smelled Domestic Cat urine, for example, you will have no difficulty recognizing the distinctive odor of wild felids, which are slightly different but sufficiently similar for identification purposes.

While smells are quite easy to distinguish, they are remarkably hard to describe. You'll just have to get out there and smell animal urine on your own. Never pass up an opportunity to refresh your memory or broaden your experience when it comes to confirmed animal scents, such as those

A Fisher mounts a stump to dribble urine as it patrols its territory.

associated with sightings, clear tracks, or associated signs that tell you which animal's scent you are smelling.

Canids and felids gravitate toward raised objects to urinate. Canids either raise a leg or squat, and often mark old stumps, rocks, low-hanging branches, and an assortment of other things. We would describe canid urine as fairly musty and dank, with the exceptions of Red and Kit Foxes, which have exceptionally strong, skunklike urines. The scent of all canid urine is strongest in winter, just before the onset of the mating season. Also, check canid urine for blood, an indication that females have come into estrus. It may be red to a slight orange, depending upon the extent of her bleeding.

Felids may raise a leg or squat, but Bobcats and Domestic Cats are more likely to squirt their urine backwards at specific objects. Bobcats seem very partial to rotten old stumps, which may absorb and hold odors longer, as well as rock overhangs and low-hanging foliage.

The position of a urine patch relative to the footprints can also help you differentiate male from female animals (Bang and Dahlstrom, 1972), at least in some animals. Deer and Elk urinate with the hind legs straddled. The urine patch of a male is between the tracks of the forefeet and the hind feet, whereas that of the female is between or behind the tracks of the hind feet. The relative position of a urine patch to scats deposited at the same time can also indicate the sex of the animal. First, use the tracks to determine the direction of travel; a urine patch in front of the scat is typical for a male, whereas a urine patch on top of or behind the scat usually indicates a female.

Like scats, urine too can accumulate at scent posts with time. Depending upon conditions, urine accumulations can even become visible to the human eye. In dry environments, accumulated urine may stain wood or rock, or evaporate to leave calcareous deposits on various elevated perches; calcareous deposits are common for woodrats and California Voles. Urine accumulations may still remain invisible—especially in moist forest conditions—but amplify the odors of each deposit, making them more difficult to overlook and ignore.

SCAT PLATES

Here is a visual presentation of the scats of many of California's animals (with an emphasis on mammals) divided into four categories: (1) pellets and small scats (plates 85–100); (2) twists, ropes, and tubes (plates 101–117); (3) loose and amorphous scats (plates 118–126); and (4) uric acid streaks and splats (plates 127–129). (Bird pellets, which resemble animal scats, are presented in plates 130–138.) Because of their variable nature, some scats are a combination of categories and thus cannot be precisely defined; however, a category had to be chosen. Note that many photos include a penny, which is a very helpful and often easily available scale in the field. A penny is exactly ¾ in. (1.9 cm) in diameter.

PLATE 85 Scats

Pellets and Small Scats

Wolf spider scats. Characteristically tiny and very hard.

American Pika: $3/32$–$1/8$ in. (0.2–0.3 cm) in diameter. Description on page 127.

Pygmy Rabbit: $1/8$–$9/32$ in. (0.3–0.7 cm) in diameter. Description on page 142.

PLATE 86 Scats

Brush Rabbit:
3/16–3/8 in.
(0.5–1 cm)
in diameter.
Description on
page 142.

Desert Cottontail:
1/4–7/16 in.
(0.6–1.1 cm)
in diameter.
Description on
page 142.

Snowshoe Hare
pellets collected
in two different
locations:
5/16–9/16 in.
(0.8–1.4 cm)
in diameter.
Description on
pages 131–132.

PLATE 87 Scats

Black-tailed Jackrabbit: $3/8–1/2$ in. (1–1.3 cm) in diameter, on the right. White-tailed Jackrabbit: $3/8–11/16$ in. (1–1.7 cm) in diameter, on the left. Descriptions on pages 135–136 and 140.

Sphinx moth frass: Up to $1/4$ in. (0.6 cm) in diameter. Note the characteristic grooved surfaces, which are characteristic of most caterpillar scats. Photo by Jonah Evans.

Deer mouse: $1/32–1/16$ in. (0.1–0.2 cm) in diameter, $3/32–7/32$ in. (0.2–0.6 cm) long. Description on page 340.

PLATE 88 Scats

Shrew: 1/16–3/16 in. (0.2–0.5 cm) in diameter, 3/16–1/2 in. (0.5–1.2 cm) long. Description on page 145.

Little Brown Myotis: 1/16–1/8 in. (0.2–0.3 cm) in diameter, 1/8–11/16 in. (0.3–1.7 cm) long. Description on page 151.

Pallid Bat: 1/16–5/32 in. (0.2–0.4 cm) in diameter, 1/8–11/16 in. (0.3–1.7 cm) long. Description on page 151.

PLATE 89 Scats

Red Tree Vole:
1/16 in. (0.2 cm)
in diameter,
5/32–3/16 in.
(0.4–0.5 cm) long.

California Vole:
1/16–1/8 in.
(0.2–0.3 cm) in
diameter, 1/8–5/16 in.
(0.4–0.8 cm) long.
Description on
page 327.

Merriam's Kangaroo
Rat: 1/16–1/8 in.
(0.2–0.3 cm) in
diameter, 1/4–1/2 in.
(0.6–1.3 cm) long.
Description on
page 311.

PLATE 90 Scats

Unknown kangaroo rat: 1/16–1/8 in. (0.2–0.3 cm) in diameter, 1/4–1/2 in. (0.6–1.3 cm) long

Long-eared Chipmunk: 3/32–5/32 in. (0.2–0.4 cm) in diameter, 1/8–9/32 in. (0.3–0.7 cm) long. Description on page 302.

White-tailed Antelope Squirrel, scats on the left and opened acorns on the right: 3/32–3/16 in. (0.2–0.5 cm) in diameter. Description on page 286.

PLATE 91 Scats

Northern Flying Squirrel: 3/32–3/16 in. 0.2–0.5 cm) in diameter, 1/8–3/8 in. (0.3–1 cm) long. Description on page 283. Photo by Chris Maser

Sparrow: 1/8–3/16 in. (0.3–0.5 cm) in diameter.

American Robin: 1/8–3/16 in. (0.3–0.5 cm) in diameter.

PLATE 92 Scats

Botta's Pocket Gopher: ⅛–³⁄₁₆ in. (0.3–0.5 cm) in diameter, ⁵⁄₁₆–⁷⁄₁₆ in. (0.8–1.1 cm) long. Description on page 320.

Bushy-tailed Woodrat: ⅛–³⁄₁₆ in. (0.3–0.5 cm) in diameter, ⁵⁄₁₆–⅝ in. (0.8–1.6 cm) long. Description on pages 334–335.

Desert Woodrat: ⅛–³⁄₁₆ in. (0.3–0.5 cm) in diameter, ¼–⁹⁄₁₆ in. (0.6–1.4 cm) long. Description on pages 334–335.

PLATE 93 Scats

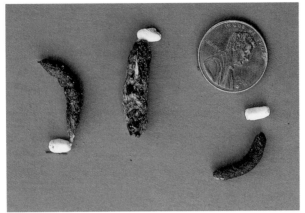

Lizard: Up to ⅜ in. (1 cm) in diameter. Note that the uric acid package is completely separate from the hard indigestible portions of prey items. This is characteristic for all the smaller lizards.

Brown Rat: ⅛–¼ in. (0.3–0.6 cm) in diameter, ¼–¹⁵⁄₁₆ in. (0.6–2.4 cm) long. Description on page 348.

Phoebe: ⅛–⁵⁄₁₆ in. (0.3–0.7 cm) in diameter. Contents are insect remains.

PLATE 94 Scats

Hairy Woodpecker: 3/16–1/2 in. (0.5–1.3 cm) in diameter. Contents are insect remains.

Common flicker fruit droppings: 1/4–3/8 in. (0.6–1 cm) in diameter.

Common flicker ant droppings: 1/4–3/8 in. (0.6–1 cm) in diameter.

PLATE 95 Scats

California Quail: ¼–½ in. (0.6–1.2 cm) in diameter.

Douglas's Squirrel: 3/16–¼ in. (0.5–0.6 cm) in diameter, 3/16–7/16 in. (0.5–1.1 cm) long. Description on pages 280–281. Photo by Liz Snair.

Golden-mantled Ground Squirrel: 3/16–5/16 in. (0.5–0.8 cm) in diameter, 3/16–½ in. (0.5–1.3 cm) long. Description on page 299.

PLATE 96 Scats

Eastern Gray Squirrel: ⅛–⁷/₃₂ in. (0.3–0.6 cm) in diameter, ⁵/₃₂–⅜ in. (0.4–1 cm) long. Description on pages 275–276.

California Ground Squirrel: ⁵/₃₂–⁵/₁₆ in. (0.4–0.8 cm) in diameter, ¼–¾ in. (0.6–1.9 cm) long. Description on page 292.

Eastern Fox Squirrel, dry scats and six fresh scats on right side: ⅛–⁵/₁₆ in. (0.3–0.8 cm) in diameter, ⁵/₃₂–⅜ in. (0.4–1 cm) long. Description on pages 275–276.

PLATE 97 Scats

Muskrat, hard pellets: 3/16–5/16 in. (0.5–0.8 cm) in diameter, 3/8–1 in. (1–2.5 cm) long. Description on page 331.

Coypu (Nutria): 3/16–1/2 in. (0.5–1.3 cm) in diameter, 3/4–1 1/2 in. (1.9–3.8 cm) long. Description on page 356.

Mourning Dove: 1/2–3/4 in. (1.3–1.9 cm) in diameter of entire scat. Scats are formed of coiled tubes to form small circles.

PLATE 98 Scats

North American
Porcupine:
¼–½ in.
(0.6–1.3 cm) in
diameter, ½–1¼ in.
(1.3–3.2 cm) long.
Description on
pages 352–353.

North American
Porcupine:
¼–½ in.
(0.6–1.3 cm)
in diameter,
½–1¼ in.
(1.3–3.2 cm) long.
Description on
pages 352–353.

Pronghorn:
³⁄₁₆–½ in.
(0.5–1.3 cm)
in diameter,
³⁄₈–¾ (1–1.9 cm)
long. Description
on page 264.

PLATE 99 Scats

Mule Deer:
³⁄₁₆–⁵⁄₈ in.
(0.5–1.6 cm) in
diameter, ½–1 ¾ in.
(1.3–4.4 cm) long.
Description on
pages 257–258.

Bighorn Sheep:
³⁄₁₆–⁵⁄₈ in.
(0.5–1.6 cm) in
diameter, ½–1 ¾ in.
(1.3–4.4 cm) long.
Description on
page 270.

Elk: ⁷⁄₁₆–¹¹⁄₁₆ in.
(1.1–1.7 cm) in
diameter, ½–1 in.
(1.3–2.5 cm) long.
Description on
page 261.

PLATE 100 Scats

Yellow-bellied Marmot: ³⁄₈–¹⁵⁄₁₆ in. (1–2.4 cm) in diameter, 1½–3½ in. (3.8–8.9 cm) long, sometimes in chains much longer. Description on pages 287–288.

American Beaver: ¾–1½ in. (1.9–3.8 cm) in diameter, 1¼–3 in. (3.2–7.6 cm) long. Description on page 305.

Ass or Feral Burro pellets: 1–1¾ in. (2.5–4.4 cm) in diameter, 2–3 in. (5.1–7.6 cm) long. Description on pages 251–252. Photo by Barry Martin.

PLATE 101 Scats

Twists, Ropes, and Tubes

Banana slug scat: $1/32–1/16$ in. (0.1–0.2 cm) in diameter. The characteristic folded coils of slugs can be found on any elevated surface.

Mole: $3/32–3/16$ in. (0.2–0.5 cm) in diameter, $3/8–1$ in. (1–2.5 cm) long. Description on page 148.

Ermine: $1/8–5/16$ in. (0.3–0.8 cm) in diameter, $3/4–2\,3/8$ in. (1.9–6 cm) long. Description on pages 223–224.

PLATE 102 Scats

Six Long-tailed Weasel scats from Seaside, California: $3/16$–$3/8$ in. (0.5–1 cm) in diameter, 1–3¼ in. (2.5–8.3 cm) long. Description on pages 223–224.

American Mink crayfish scat: ¼–$7/16$ in. (0.6–1.1 cm) in diameter, 1–4 in. (2.5–10.2 cm) long. Description on page 228.

Two American Mink meat scats with miniature disposable lighter: ¼–$7/16$ in. (0.6–1.1 cm) in diameter, 1–4 in. (2.5–10.2 cm) long. Description on page 228.

PLATE 103 Scats

Pileated Woodpecker: $^3/_{16}$–$^1/_2$ in. (0.5–1.3 cm) in diameter. Composed of carpenter ant remains.

Ruffed Grouse: $^1/_4$–$^3/_8$ in. (0.6–1 cm) in diameter.

Five American Marten fruit scats: $^3/_{16}$–$^5/_8$ in. (0.5–1.6 cm) in diameter, 2–5 in. (5.1–12.7 cm) long. Description on page 214.

PLATE 104 Scats

Five American
Marten meat scats:
³/₁₆–⁵/₈ in.
(0.5–1.6 cm) in
diameter, 2–5 in.
(5.1–12.7 cm) long.
Description on
page 214.

Ringtail: ³/₁₆–⁵/₈ in.
(0.5–1.6 cm) in
diameter, 2½–5 in.
(6.4–12.7 cm) long.
Description on
pages 242–243.

Ringtail: ³/₁₆–⁵/₈ in.
(0.5–1.6 cm) in
diameter, 2½–5 in.
(6.4–12.7 cm) long.
Description on
pages 242–243.
Photo by
Jonah Evans.

PLATE 105 Scats

Three Kit Fox scats: 3/16–5/8 in. (0.5–1.6 cm) in diameter, 2–4½ in. (7.6–11.4 cm) long. Description on pages 184–185.

Four Western Toad scats: ¼–5/8 in. (0.6–1.6 cm) in diameter. Toad scats are large relative to the size of the animal and typically slightly curved. Because they are composed of insect parts, people often mistake them for skunk scats.

Three Western Diamondback Rattlesnake scats: ¼–7/8 in. (0.6–2.2 cm) in diameter. Snake scats are twisted ropes with thick, sticky uric secretions coating some portion of the defecation.

PLATE 106 Scats

Unknown snake species (likely a Gopher Snake in this case): ¼–⅞ in. (0.6–2.2 cm) in diameter.

Fisher apple scat: ³⁄₁₆–¾ in. (0.5–1.9 cm) in diameter, 2–6 in. (5.1–15.2 cm) long. Description on page 219.

Fisher meat scat: ³⁄₁₆–¾ in. (0.5–1.9 cm) in diameter, 2–6 in. (5.1–15.2 cm) long. Description on page 219.

PLATE 107 Scats

Male Wild Turkey: 3/8–5/8 in. (1–1.6 cm) in diameter. Tom scats are straighter or J-shaped.

Canada Goose: 7/16–5/8 in. (1.1–1.6 cm) in diameter.

American Crow: ¼–1 in. (0.6–2.5 cm) in diameter.

PLATE 108 Scats

(Californian) Desert Tortoise: ¼–1 in. (0.6–2.5 cm) in diameter.

Common Gray Fox: ⅜–¾ in. (1–1.9 cm) in diameter, 3–5½ in. (7.6–14 cm) long. Description on page 181.

Two Common Gray Fox scats—meat on the left and fruit on the right: ⅜–¾ in. (1–1.9 cm) in diameter, 3–5½ in. (7.6–14 cm) long. Description on page 181.

PLATE 109 Scats

Red Fox: $5/16$–$3/4$ in. (0.8–1.9 cm) in diameter, 3–6 in. (7.6–15.2 cm) long. Description on pages 188–189.

Domestic Cat: $3/8$–$7/8$ in. (1–2.2 cm) in diameter. Description on page 153.

Striped Skunk: $3/8$–$7/8$ in. (1–2.2 cm) in diameter, 2–5 in. (5.1–12.7 cm) long. Description on page 236.

PLATE 110 Scats

Striped Skunk: ³⁄₈–⁷⁄₈ in. (1–2.2 cm) in diameter, 2–5 in. (5.1–12.7 cm) long. Description on page 236.

Female Wild Turkey: ¼–1¼ in. (0.6–3.2 cm) in diameter of entire scat.

Roadrunner: ¾–1¼ in. (1.9–3.2 cm) in diameter of entire scat. The thick, viscous white cap is characteristic of this species.

PLATE 111 Scats

North American River Otter crayfish scats: ⅜–1 in. (1–2.5 cm) in diameter, 3–6 in. (7.6–15.2 cm) long. Description on pages 205–206.

Two North American River Otter scats, one fresh, one old: ⅜–1 in. (1–2.5 cm) in diameter, 3–6 in. (7.6–15.2 cm) long. Description on pages 205–206.

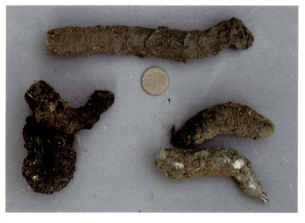

Three American badger scats from Fort Ord, California: ⅜–¹⁵⁄₁₆ in. (1–2.4 cm) in diameter, 3–6 in. (7.6–15.2 cm) long. Description on page 232.

PLATE 112 Scats

Wolverine:
³⁄₁₆–1 in. (1–2.5 cm)
in diameter, 3–8 in.
(7.6–20.3 cm) long.
Description on
page 211.

Virginia Opossum:
³⁄₈–1 ¹⁄₈ in.
(1–2.9 cm) in
diameter, 1–4 ½ in.
(2.5–11.4 cm) long.
Description on
page 123.

Northern Raccoon:
⁵⁄₁₆–1 ³⁄₁₆ in.
(0.8–3 cm) in
diameter, 3½–7 in.
(8.9–17.8 cm) long.
Description on
pages 246–247.

PLATE 113 Scats

Northern Raccoon: 5/16–1 3/16 in. (0.8–3 cm) in diameter, 3½–7 in. (8.9–17.8 cm) long. Description on pages 246–247.

Bobcat: 7/16–1 in. (1.1–2.5 cm) in diameter, 3–9 in. (7.6–22.9 cm) long. Description on page 257.

Bobcat: 7/16–1 in. (1.1–2.5 cm) in diameter, 3–9 in. (7.6–22.9 cm) long. Description on page 157

PLATE 114 Scats

Coyote juniper and hide scat: ⅜–1⅜ in. (1–3.5 cm) in diameter. Description on pages 171–172.

Domestic Dog: All sizes. Description on page 169. Photo by Laura Feinstein.

Coyote grass scat: ⅜–1⅜ in. (1–3.5 cm) in diameter. Description on pages 171–172.

PLATE 115 Scats

American Beaver: ¾–1½ in. (1.9–3.8 cm) in diameter, 1¼–3 in. (3.2–7.6 cm) long. Description on page 305.

Cougar (Mountain Lion): ¾–1⅝ in. (1.9–4.1 cm) in diameter, 6½–17 in. (16.5–43.2 cm) long. Description on page 163.

Cougar (Mountain Lion): ¾–1⅝ in. (1.9–4.1 cm) in diameter, 6½–17 in. (16.5–43.2 cm) long. Description on page 163.

PLATE 116 Scats

Sea Otter:
1–1 ½ in.
(2.5–3.8 cm)
in diameter.
Description on
page 202. Photo
by Max Allen.

California Sealion:
1–2 in.
(2.5–5.1 cm)
in diameter.
Description on
page 199.

Wolf: ½–1 ⅞ in.
(1.3–4.8 cm) in
diameter, 6–17 in.
(15.2–43.2 cm)
long. Description
on page 177.

PLATE 117 Scats

Wild Boar: 1–2 in. (2.5–5.1 cm) in diameter, 4–14 in. (10.2–35.6 cm) long. Description on pages 253–254.

Wild Boar: 1–2 in. (2.5–5.1 cm) in diameter, 4–14 in. (10.2–35.6 cm) long. Description on pages 253–254.

American Black Bear grass scat: 1¼–2½ in. (3.2–6.4 cm) in diameter, 5–12 in. (12.7–30.5 cm) long. Description on page 194.

PLATE 118 Scats

Loose and Amorphous Scats

American Pika. Description on page 127.

Muskrat, soft: ⅜–1 in. (1–2.5 cm) in diameter, ⅞–1¾ in. (2.2–4.4 cm) long. Description on page 331.

Ruffed Grouse: soft cecal droppings, which have followed and landed atop a pile of typical grouse droppings.

PLATE 119 Scats

Cecal dropping of a pheasant.

North American River Otter. Description on pages 205–206.

Sea Otter. Description on page 202. Photo by Max Allen.

PLATE 120 Scats

Pronghorn, clumped pellets. Description on page 264.

Mule Deer, clumped. Description on pages 257–258.

Mule Deer. Description on pages 257–258.

PLATE 121 Scats

Wild Boar.
Description on
pages 253–254.

Bighorn Sheep.
Description on
page 270.

Elk, medium-soft.
Description on
page 261.

PLATE 122 Scats

Elk, soft. Description on page 261.

Coyote apple scat. Description on pages 171–172.

Coyote soft meat scat. Description on pages 171–172.

PLATE 123 Scats

Cougar (Mountain Lion), adult female, soft scat composed of internal organs at a fresh kill. Description on page 163.

Harbor Seal. Description on page 201.

California Sealion. Description on page 199.

PLATE 124 Scats

American Black
Bear berry scat.
Description on
page 194.

American Black
Bear apple scat.
Description on
page 194.

American Black
Bear manzanita
scat. Description
on page 194.

PLATE 125 Scats

American Black Bear acorn scat. Description on page 194.

Horse. Description on pages 251–252.

American Bison, chips: Individual chips range in diameter from 3–4 ½ in. (7.6–11.4 cm). Description on page 267.

PLATE 126 Scats

American Bison, patties: 10–16 in. (25.4–40.6 cm) in diameter. Description on page 267.

PLATE 127 Scats

Uric Acid Streaks and Splats

Tracks and splat from a Great Blue Heron. Splats can measure up to 18 in. (45.7 cm) across.

The characteristic splat of a white pelican.

Accumulated uric acid splats of gulls at a regular perch.

PLATE 128 Scats

Uric acid streaks created by an Osprey while perched and feeding on the ground.

Paired uric acid streaks of a loafing Double-crested Cormorant.

The large uric acid streak made by a Bald Eagle while it was perched on the log.

PLATE 129 Scats

Accumulated streaks of Great Horned Owls under a bridge adjacent to their nest.

BIRD PELLETS
Mark Elbroch

Splats and pellets accumulating under the regular roost of a Great Horned Owl.

Pellets are compressed masses of indigestible food remains regurgitated by birds that resemble some scats of mammals; they are composed of hair, bones, feather scraps, undigested hulls of mast crops, fish scales, insect carapaces, shells of bivalves, and numerous other odds and ends, including plastics and garbage. Owl pellets are notorious, yet grebes, cormorants, vultures, hawks, eagles, herons, rails, gulls, terns, crows, jays, thrushes, kinglets, flycatchers, shrikes, and warblers are also capable of producing pellets (Leahy 1982). However, pellets of small birds are very difficult to find, as well as those composed of anything but garbage, fur, or bone, because they deteriorate so quickly in nature.

Like mammals, birds have a simple esophagus, which in birds is often called the gullet. It runs from the mouth to their two-part stomach. In contrast to mammals, many birds can temporarily expand their gullets to store food internally, and their two-part stomachs function much as teeth do in mammals. Lacking teeth, birds need to either tear foods into smaller pieces to fit them into their mouths, as do hawks and owls, or crush the food item to break it into smaller pieces, as do many seed eaters, or swallow their food whole, like herons. Food items then move down the gullet and into the first part of the stomach, called the proventriculus, where it is liberally doused in digestive juices. Then the food moves into the second compartment of the stomach (often called the gizzard), where grit and sand eaten by the birds, as well as the muscular rough, sandpaper-like walls begin to grind foods down. Any food items that the gizzard cannot break down and pass to the small intestine for further digestion, are compressed into a pellet and then moved back into the proventriculus. There the pellet remains until the bird goes next to feed, and muscle contractions move the pellet up the esophagus and out the bird's mouth. Watching a bird eject a pellet can be jarring, because it sometimes appears as if they are choking, but it does not harm the bird; in fact the mucus-covered pellets scour the gullet clean as they move up and out, thus preparing the passage for future meals.

Finding Bird Pellets

Raptor pellets can easily be collected at their nests and regular roosting sites. For owls, these are often large, prominent trees near excellent hunting areas with substantial foliage to shield them during the daylight hours. Telephone poles and fences are excellent sites for falcons, hawks, and allies. Kingfisher pellets also accumulate at their regular hunting perches; you just have to find one that isn't above water.

The pellets of cormorants, gulls, and terns can be found at loafing areas, where great flocks of birds rest on shore. Sometimes flocks are of a single species, making pellet identification easy, but at other times they are mixed flocks and tracks must be studied so as to distinguish between them. Note that numerous loafing areas are only exposed at low tide, and thus the area is swept clean of scats and pellets twice daily.

Pellets of numerous waders are easy to find if you follow their trails long enough. Sanderlings and others produce pellets as they forage, barely breaking their stride as they continue to probe for more food.

Compare the bones in similarly sized Red-tailed Hawk (the tiny pile) and Great Horned Owl pellets.

Differentiating Hawk and Owl Pellets

The pellets of similar-sized hawks and owls can be differentiated by contents. Hawks tear their prey into smaller pieces and generally avoid swallowing larger bones. They also have stronger stomach acids than owls. In contrast, owls swallow prey whole or tear them into larger pieces than hawks do. They also have weaker stomach acids, so entire bone fragments and small skulls appear in their pellets, beautifully preserved. Compare the contents of two similar-sized pellets in the photograph above.

Identifying Pellets

Differentiating the pellets of various species in the field can be a difficult task. Pellets associated with tracks, regular roosts, and nests are certainly the most dependably identified.

The size of pellets produced by a single bird varies dramatically depending on diet and the size of its meal. Pellets that contain only insect parts are significantly smaller than those of fur and bone, even when produced by the same bird. Therefore, there is substantial overlap across species, leading to such large measurement parameters that the numbers lose some value to the tracker and naturalist. Certainly birds have an upper dimension that a pellet cannot physically exceed, and in our research the pellet's width is the most useful measurement to take and compare.

What follows is a visual presentation of select pellets of California birds, arranged from those with the smallest width to those with the largest. Data and visual aids for several additional species can be found in Elbroch and Marks (2001).

BIRD PELLET PLATES

PLATE 130 Bird Pellets

Sanderling (*Calidris alba*): ¼–⅜ in. (0.6–0.9 cm) wide.

American Kestrel (*Falco sparverius*), insects: ⁵⁄₁₆–⁷⁄₁₆ in. (0.8–1.1 cm) wide.

Loggerhead Shrike (*Lanius ludovicianus*): ⁵⁄₁₆–½ in. (0.8–1.2 cm) wide.

PLATE 131 Bird Pellets

Burrowing Owl
(*Athene cunicularia*):
³⁄₈–⁹⁄₁₆ in.
(0.9–1.4 cm) wide.

Black-billed Magpie
(*Pica hudsonia*):
³⁄₈–⁹⁄₁₆ in.
(0.9–1.4 cm) wide.

Black Oystercatcher
(*Haematopus
bachmani*): ⁵⁄₈ in.
(1.6 cm) wide.

PLATE 132 Bird Pellets

American Crow (*Corvus brachyrhynchos*): ⁵/₁₆–¾ in. (0.8–1.9 cm) wide.

Willet (*Catoptrophorus semipalmatus*): ³/₈–⁵/₈ in. (0.9–1.6 cm) wide. Photo by Jonah Evans.

Clark's Nutcracker (*Nucifraga columbiana*): ⁵/₁₆–¾ in. (0.8–1.9 cm) wide.

PLATE 133 Bird Pellets

Northern Saw-whet Owl (*Aegolius acadicus*): $7/16$–$11/16$ in. (0.9–1.6 cm) wide.

Black Skimmer (*Rynchops niger*): $1/2$–$5/8$ in. (1.2–1.6 cm) wide.

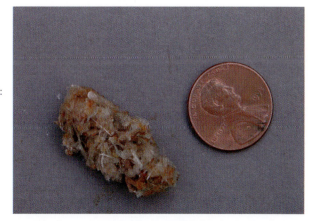

Belted Kingfisher (*Megaceryle alcyon*): $1/2$–$3/4$ in. (1.2–1.9 cm) wide.

PLATE 134 Bird Pellets

Broad-winged Hawk
(*Buteo platypterus*):
⁹⁄₁₆–¾ in.
(1.4–1.9 cm) wide.

Short-eared Owl
(*Asio flammeus*):
½–⅞ in.
(1.2–2.2 cm) wide.

Barred Owl
(*Strix varia*):
⅝–¾ in.
(1.6–1.9 cm) wide.

PLATE 135 Bird Pellets

Heermann's Gull
(*Larus heermanni*):
⅝–⅞ in.
(1.6–2.2 cm)

Red-tailed Hawk
(*Buteo jamaicensis*):
⁹⁄₁₆–¹⁵⁄₁₆ in.
(1.4–2.4 cm) wide.

Common Raven
(*Corvus corax*):
⁹⁄₁₆–¹⁵⁄₁₆ in.
(1.4–2.4 cm) wide.

PLATE 136 Bird Pellets

Common Raven
(*Corvus corax*):
⁹⁄₁₆–¹⁵⁄₁₆ in.
(1.4–2.4 cm) wide.
These pellets are
composed of trash.

Turkey Vulture
(*Cathartes aura*):
¹³⁄₁₆–1 in.
(2–2.5 cm) wide.

Barn Owl (*Tyto alba*):
¹¹⁄₁₆–1 ⅛
(1.7–2.8 cm) wide.

PLATE 137 Bird Pellets

Herring Gull
(*Larus argentatus*):
¾–1 ³/₁₆ in.
(1.9–3 cm) wide.

Black-shouldered
Kite (*Elanus
axillaris*): ⁷/₈–1 ¼ in.
(2.2–3.1 cm) wide.

Great Horned Owl
(*Bubo virginianus*):
⁹/₁₆–1 ½ in.
(1.4–3.8 cm) wide.

PLATE 138 Bird Pellets

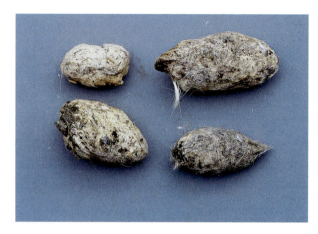

Golden Eagle
(*Aquila chrysaetos*):
$^{11}/_{16}$–1 ½ in.
(1.7–3.8 cm) wide.

MAMMAL SPECIES ACCOUNTS

Mark Elbroch, Jonah Evans, and Michael Kresky

THE FOLLOWING MAMMALIAN species accounts are organized following the taxonomical order found in *Mammal Species of the World* (Wilson and Reeder 2005) and provide more detailed information about each species' tracks, trails, scats, and other scent-marking behaviors. Our understanding of phylogeny and taxonomy continues to expand and change, resulting in a better understanding of how mammals are related to each other and in what order they have evolved. The first mammal to appear below, the Virginia Opossum, *Didelphis virginiana*, is considered the oldest mammal in California in evolutionary terms—meaning that the opossum as we know it today reached its current form before all others. In truth, groups of animals appeared before others, and thus speciation occurs simultaneously on many fronts, on different branches of the "phylogenetic tree." Even more exciting, we now have mountains of irrefutable evidence that evolution is continuous and proceeds much faster than we imagined it might just fifty years ago. (For an easy-to-read introduction to the subject, consider Jonathan Weiner's [1995] *The Beak of the Finch: A Story of Evolution in Our Time*). Indeed, evolution occurs from one generation to the next. Thus, none of the species presented here are static, but each is continually changing. This is an especially important lesson with regard to tracking wildlife. We must constantly be flexible in our interpretations, because animals constantly surprise us.

A benefit of keeping mammals in family groups is in training our minds to see how species are related to each other, and how some behavioral and morphological characteristics are shared by many species. Learn to recognize tracks as members of a group of sign belonging to an order or family; this will aid in your tracking as you travel beyond California.

ORDER DIDELPHIMORPHIA

Opossums: Family Didelphidae

VIRGINIA OPOSSUM *Didelphis virginiana*

TRACKS: Front: 1–2⁵/₁₆ in. long (2.5–7.5 cm) × 1¼–2½ in. wide (3.2–6.3 cm)

Small to medium. Plantigrade. Symmetrical. Five toes, starlike and radiating from the metacarpal region like rays of a rising sun. Five metacarpal pads and an additional carpal pad. Claws may or may not register.

Hind: 1³/₁₆–2¾ in. long (3–7 cm) × 1½–3 in. wide (3.8–7.6 cm)

Small to medium. Plantigrade. Highly asymmetrical and distinctive. Five toes. Toe 1, the opposable thumb, sits across and separate from the remaining digits, pointing into the center of the trail. Toes 2, 3, and 4 tend to register together as a tight group, and digit 5 tends to parallel them but registers as slightly separated. Four prominent metatarsal pads (pads for toes 3 and 4 are fused). Toe 1 lacks a claw, and the remaining claws may or may not register.

TRAILS:

Walk:	Strides: 4½–9 in. (11.4–22.9 cm)	
	Trail widths: 3¾–5 in. (9.5–12.7 cm)	
Trot:	Strides: 10–15 in. (25.4–38.1 cm)	
	Trail widths: 2½–4 in. (6.3–10.2 cm)	
Bound:	Strides: 13–19 in. (33–48.3 cm)	
	Trail widths: 5–7½ in. (12.7–19.1 cm)	
	Group lengths: 9–11½ in. (22.9–29.2 cm)	

NOTES:

· Opossums are exotic species, and were introduced to California in the earliest 1900s. The first reported animal caught in the state was in 1906, and then numerous reports started in coastal and southern California near 1915 (Grinnell, Dixon, and Linsdale 1937), where today they are now numerous.

· Walk while foraging or under cover, but often switch to a trot to cross open ground or travel beaches and trails more quickly. These two gaits leave very similar trail patterns and could be confused.

· Often the hind track lands partially on top of and slightly behind the front track, creating a confusing jumble of toes and pads. Look closely to differentiate the several front toes that project outward from the print of the hind track.

· Walking opossums occasionally drag their tails, but trotting animals (or those moving faster) hold their tail straight out behind them, keeping it well clear of the ground. The meandering marks left by the trail could easily be confused for a snake if the footprints that accompany them are overlooked.

· Opossums use a bound to escape or chase prey, although in encounters with predators they often "play dead" instead of attempting to run away.

· They are capable climbers and swimmers.

· They use holes and cavities, including drains, culverts, and basements, to rest and spend periods of extreme cold. Opossums do not venture out often in deep snow, because their feet and tails are prone to frostbite.

SCAT: ³/₈–1¹/₈ in. (1–2.9 cm) diameter, 1–4½ in. (2.5–11.4 cm) long

Opossum scat is highly variable, depending upon diet, and most often soft or loose. Opossums deposit scats at random along their travel routes. Captive opossum scat often contains numerous body hairs that they ingest while grooming themselves with their tongues (McManus 1970).

URINE AND OTHER SCENT-MARKING BEHAVIORS: Opossums scent-mark trees and other standing structures with saliva and glands on their cheeks (McManus 1970).

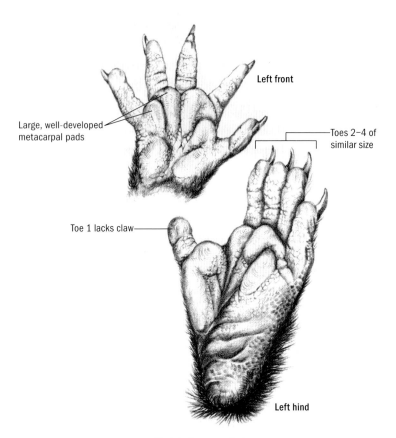

Large, well-developed metacarpal pads

Toes 2–4 of similar size

Left front

Toe 1 lacks claw

Left hind

Virginia Opossum feet.

Trotting Virginia Opossum.

Walking Virginia Opossum.

Front and hind tracks of a Virginia Opossum.

The trail made by a trotting Virginia Opossum.

ORDER LAGOMORPHA

Pikas: Family Ochotonidae

AMERICAN PIKA   *Ochotona princeps*

Plate 15
Plates 85 and 118

TRACKS: Front: $^{11}/_{16}$–$^7/_8$ in. long (1.7–2.2 cm) × $^{11}/_{16}$–$^7/_8$ in. wide (1.7–2.2 cm)

Small. Digitigrade. Asymmetrical. Round tracks with bulbous digital pads. Five toes. Toe 1 is reduced but proportionately larger than in rabbit tracks. Sole is furred. Claws may or may not register. Front tracks similar in size to hind tracks.

Hind: ¾–1$^1/_8$ in. long (1.9– 2.9 cm) × ¾–1 in. wide (1.9–2.5 cm)

Small. Digitigrade. Asymmetrical. Four toes. Roundish track, with a furred and indistinct sole. Claws may or may not register.

TRAILS: Bound: Strides: 2¼–15 in. (5.7–38.1 cm)
            Trail widths: 2–3 in. (5.1–7.6 cm)
            Group lengths: 2–4$^7/_8$ in. (5.1–12.4 cm)

NOTES:

• Bound when traveling to and from foraging areas, often using the same path back and forth.

• Alpine creatures, most often associated with rock jumbles and scree. In some areas, they will also be found in higher elevation forests, again seeking shelter in rock jumbles.

• In deep-snow conditions pikas tend to move below the surface, forming runs, latrines, and surviving on hay piles created during the summer.

SCAT: Pellets $^3/_{32}$–$^1/_8$ in. (0.2–0.3 cm) diameter

Pikas produce two types of scats: the hard, brown, round pellets that are frequently encountered in the field, and cecal pellets that look like brown or green toothpaste squeezed from a tube. Pikas post their hard scats in high-traffic areas, including burrow entrances and trail junctions; they also defecate in large latrines, where pellets may accumulate in the hundreds, if not thousands. Cecal pellets are either reingested or deposited and stored in their hay piles for later use (Smith and Weston 1990).

URINE AND OTHER SCENT-MARKING BEHAVIORS: Both male and female Pikas scent-mark with glands found on their cheeks, which they rub on rocks and other projecting surfaces throughout their home range (Svendsen 1979). Scent-marking increases in areas where home ranges of males and females overlap, and suggests that scent-marking may play a dual role. Scents probably maintain territorial boundaries but also probably advertise the sexual status and health of an individual to prospective mates (Svendsen 1979; Stewart, Scudder, and Southwick 1982).

Pikas also urinate regularly at the entrance to regular trails within rock talus and near burrows. Where they urinate repeatedly, nitrogen accumulates, creating perfect habitat for the elegant sunburst lichen (*Xanthoria elegans*) (Laws 2007). You can even use this flat orange lichen that grows on sun-exposed rocks to locate Pika latrines, such is the lichen's dependence upon mammals and birds to create suitable circumstances for its survival.

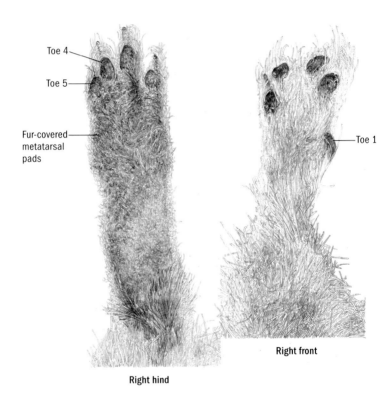

Toe 4

Toe 5

Fur-covered metatarsal pads

Toe 1

Right front

Right hind

American Pika feet.

Bounding trail of
American Pika.

The trail of a bounding American Pika on old
snow pack.

The large winter latrine of an American Pika.

White calcareous deposits created by repeated urinations and pellets mark a well-used entrance of an American Pika burrow in the talus. Photo by Jonah Evans.

Rabbits and Hares: Family Leporidae

SNOWSHOE HARE   *Lepus americanus*

TRACKS: Front: 1⅞–3 in. long (4.8–7.6 cm) × 1⅛–2¼ in. wide (2.9–5.7 cm)

Medium. Digitigrade. Highly asymmetrical and often pointy, as a result of a single toe leading all the others. Five toes. Toe 1 is greatly reduced, but still clawed and occasionally registers in tracks. Digital pads are absent, but the impressions of fur-covered toes are often evident in tracks. Palm is furred and often indistinct. Claws may or may not register, but are often evident in good substrate. Front tracks significantly smaller than hind tracks and significantly larger than the front tracks of cottontails.

Hind: 3¼–6 in. long (8.3–15.2 cm) × 1⅝–5 in. wide (4.1–12.7 cm)

Medium to very large. Digitigrade. Asymmetrical. Four toes. Digital pads and sole are furred and often indistinct. Large furred heel tends to register. Claws may or may not register. Tracks rounder than other rabbits. When splayed, tracks can be massive, more symmetrical, and easily confused with dog tracks. In hard substrates, claws are the predominant or only visible sign of the track.

TRAILS: Bound: Strides: 8–72 in. (20.3–182.9 cm) up to 10 ft (304.8 cm)
                      Trail widths: 3¾–10 in. (9.5–25.4 cm)
                      Group lengths: 8–30 in. (20.3–76.2 cm)

NOTES:

• Although hare trails are often found under cover, they will also cross forest gaps and fields. They inhabit northern woodlands and higher altitudes in California.

• Hares do not use burrows but rest in forms; however, they will dig snow caves when the temperatures are very low, or use existing protective pockets within snow-covered vegetation.

• Tend to keep their hind feet parallel when bounding, except when moving at full speed, when their patterns more resemble those of jackrabbits. They are the only North American hare that regularly bound with the hind feet landing side by side.

SCAT: Pellets ⁵⁄₁₆–⁹⁄₁₆ in. (0.8–1.4 cm) diameter

All rabbits and hares engage in coprophagy, the ingesting of their feces to better absorb plant nutrients. The first scat is very hard to find in the field, but the pellets that are produced after the second ingestion are plentiful. The scats of all leporids look similar—round pellets or disks. Rabbit scats tend to look like a squashed sphere but are occasionally teardrop-shaped as well. Pellets are composed entirely of plant material; in winter the woody material is obvious and looks like sawdust.

Rabbits release one pellet at a time, instead of the shotgun approach used by deer. Therefore, an accumulation of rabbit pellets means that either the animal remained in one place for an extended period or it has returned there one or more times.

URINE AND OTHER SCENT-MARKING BEHAVIORS: Urine is often orange to red in color, and is easily mistaken for blood.

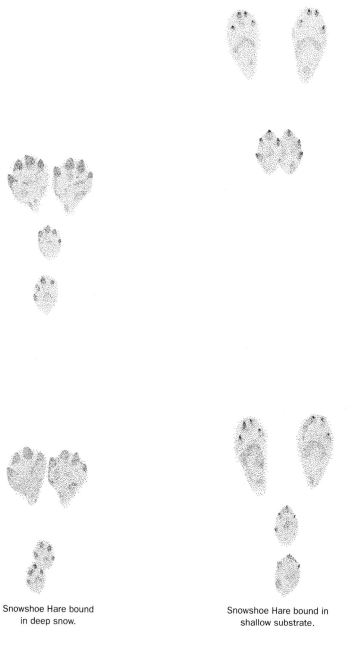

Snowshoe Hare bound
in deep snow.

Snowshoe Hare bound in
shallow substrate.

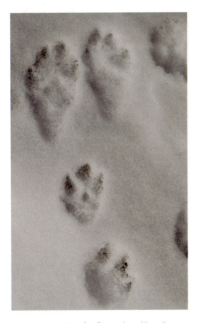

The tight tracks of a Snowshoe Hare in shallow snow. Note the size of the front tracks.

The splayed tracks of a Snowshoe Hare in snow.

Snowshoe Hare scats have accumulated while foraging over a long duration. They are eating the cambium off of this aspen branch.

BLACK-TAILED JACKRABBIT   *Lepus californicus*

TRACKS: Front: 1⁷⁄₈–2½ in. long (4.8–6.4 cm) × 1¼–1¾ in. wide (3.2–4.4 cm)

Small to medium. Digitigrade. Highly asymmetrical and often pointy, as a result of a single toe leading all the others. Five toes. Toe 1 is greatly reduced, but still clawed and occasionally registers in tracks. Digital pads are absent, but the impressions of fur-covered toes are often evident in tracks. Palm is furred and often indistinct. Claws may or may not register, but are often evident in good substrate. Front tracks smaller than hind tracks, but in shallow substrates front and hind tracks can be similar dimensions.

Hind: 2½–5⁵⁄₈ in. long (6.4–14.3 cm) × 1³⁄₈–2½ in. wide (3.5–6.4 cm)

Medium to large. Digitigrade. Asymmetrical. Four toes. Furred sole is often indistinct. Large furred heel registers often. Claws may or may not register, but are often evident in good substrate. Tracks pointy, unless splayed, when the anterior edges appear rounder. In hard substrates claws are the predominant or only visible sign of the track.

TRAILS: Bound: Strides: 7–93 in. (17.8–236.2 cm) up to 20 ft (609.6 cm)
Track width: 3–6 in. (7.6–15.2 cm)
Group lengths: 11–24 in. (27.9–61 cm)

NOTES:

- It can at times be difficult to differentiate small jackrabbits from large cottontails. The width of the front and hind feet can be a useful characteristic as long as you account for splay—meaning that you should avoid comparing a splayed cottontail track with a tight jackrabbit track, because then the measurements can create greater confusion. Also compare the length of the hind foot if it registers in its entirety, because those of jackrabbits are significantly longer than the hind feet of cottontails.

- Jackrabbits use a modified bound, in which one hind foot touches down before and in front of the other; this is technically a gallop when the hind feet do not land simultaneously.

- Sometimes alternate long, fast bounds or gallops with a more vertical jump, to allow a better view of their environment. Ernest Thompson Seton called them "spy jumps" (Seton 1958). Murie and Elbroch (2005) report that this species also walks on occasion.

- This species is considerably smaller than the White-tailed Jackrabbit, and in much of dry California is the only jackrabbit present.

SCAT: Pellets ³⁄₈–½ in. (1–1.3 cm) diameter

Black-tailed Jackrabbits defecate on average 545 pellets per day, and the size of their pellets increases with increased consumption of forage (Arnold and Reynolds 1943). All rabbits and hares engage in coprophagy, the ingesting of their feces to better absorb plant nutrients. Scats of all leporids look similar—round pellets or disks.

Rabbit scats tend to look like squashed spheres but are occasionally teardrop-shaped as well. Pellets are composed entirely of plant material; in winter the woody material is obvious and looks like sawdust.

Rabbits release one pellet at a time, rather than the shotgun approach used by deer. Therefore, an accumulation of rabbit pellets indicates that either the animal remained in one place for a long time or it has returned there one or more times.

URINE AND OTHER SCENT-MARKING BEHAVIORS: They have two rectal glands that secrete substances that appear yellow or orange when dry and hardened. Jackrabbit urine leaves colorful stains in areas where they forage, and the color may be a combination of diet and secretions from these glands.

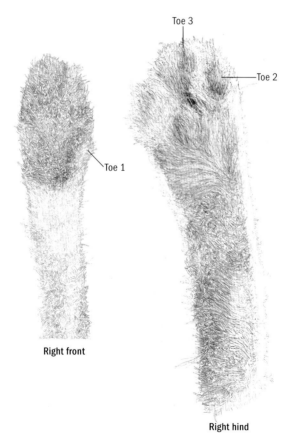

Black-tailed Jackrabbit feet.

Black-tailed
Jackrabbit lope.

Black-tailed
Jackrabbit gallop.

The galloping trail of a Black-tailed Jackrabbit. The second group of four tracks is the harder landing after the animal completed a sky jump.

The hind tracks of a Black-tailed Jackrabbit in mud. Photo by Jonah Evans.

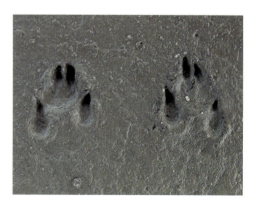

Jackrabbit urine.

WHITE-TAILED JACKRABBIT

Lepus townsendi

Plate 30
Plate 87

TRACKS: Front: 2¹/₈–3¾ in. long (5.4–9.5 cm) × 1½–2⁵/₈ in. wide (3.8–6.7 cm)

Medium. Digitigrade. Highly asymmetrical and often pointy, as a result of a single toe leading all the others. Five toes. Toe 1 is greatly reduced, but still clawed and occasionally registers in tracks. Digital pads are absent, but the impressions of fur-covered toes are often evident in tracks. Palm is furred and often indistinct. Claws may or may not register, but are often evident in good substrate. Front tracks smaller than hind tracks, but in shallow substrates the two are often similar dimensions.

Hind: 2½–6¾ in. long (6.4–17.1 cm) × 1⁵/₈–3³/₈ in. wide (4.1–8.6 cm)

Medium to large. Digitigrade. Asymmetrical. Four toes. Furred sole is often indistinct. Large furred heel registers often. Claws may or may not register, but are often evident in good substrate. Tracks pointy, unless splayed, when the anterior edges appear rounder. In some substrates claws are the predominant or only visible sign of the track.

TRAILS: Bound: Strides: 9–72 in. (22.9–182.9 cm) up to 19 ft (579.1 cm)
Trail widths: 4½–9 in. (11.4–22.9 cm)
Group lengths: 20–45 in. (50.8–114.3 cm)

NOTES:

• Jackrabbits use a modified bound, in which one hind foot touches down before and in front of the other; this is technically a gallop when the hind feet do not land simultaneously.

• The largest rabbit within its range, and inhabits open habitats, including high-elevation deserts, open forests, and mountain passes.

SCAT: *See also* the Snowshoe Hare account for a general description of rabbit scat.

Pellets ³⁄₈–¹¹⁄₁₆ in. (1–1.7 cm) diameter

White-tailed Jackrabbit gallop.

The trail of a bounding White-tailed Jackrabbit.

SMALL RABBITS   *Sylvilagus* and *Brachylagus* spp.

Plate 20

Plates 85 and 86

TRACKS: Front: $^7/_8$–$1^7/_8$ in. long (2.2–4.8 cm) × $^5/_8$–$1^3/_8$ in. wide (1.6–3.5 cm)

Small. Digitigrade. Highly asymmetrical and often pointy, because a single toe is leading all the others. Five toes. Toe 1 is greatly reduced, but still clawed and occasionally registers in tracks. Digital pads are absent, but tracks often reveal the impressions of fur-covered toes. Sole is furred and often indistinct. Claws may or may not register, but are often evident in good substrate. Front tracks significantly smaller than hind tracks.

Hind: $1^1/_4$–$3^1/_4$ in. long (3.2–8.3 cm) × $^3/_4$–$1^{13}/_{16}$ in. wide (1.9–4.6 cm)

Small to medium. Digitigrade. Asymmetrical. Four toes. Furred sole is often indistinct. Large furred heel registers often. Claws may or may not register, but are often evident in good substrate. Tracks pointy, unless splayed, when the anterior edges appear rounder. In some substrates, claws are the predominant or only visible sign of the track.

TRAILS: Bound: Strides: 4–32 in. (10.2–81.3 cm)
Trail widths: 2–5 in. (5.1–12.7 cm)
Group lengths: 4–18 in. (10.2–45.7 cm)

NOTES:

- The Brush Rabbit and Desert Cottontail are the two common small rabbits in California, and their tracks are remarkably similar and difficult to differentiate. Luckily, the two species can generally be distinguished by distribution (Brush Rabbits are primarily coastal, and in the foothills east of the Sacramento Valley, and Desert Cottontails are primarily in the arid portions in the lower three-quarters of California). In addition, the tracks of Brush Rabbits are boxier, and the central toes appear shorter than in cottontails.

- Pygmy Rabbits have a tiny distribution in eastern California and are the only rabbits that dig their own burrows. Desert Cottontails often seek shelter in existing burrows, such as those dug by California Ground Squirrels or Badgers; they do not dig their own burrows.

- Cottontails and Brush Rabbits bound when foraging and traveling, and prefer to stay in or near thick cover.

- Cottontails and Brush Rabbits bound in deep snow, and the drag marks created by the front feet follow the midline of the trail. Drag marks in squirrels follow the outer edge of the trail.

- It can at times be difficult to differentiate small jackrabbits from large cottontails. *See* comments under Black-tailed Jackrabbit above.

- The riparian Brush Rabbit near San Francisco is unique in that it often climbs shrubs to escape rising tides.

SCAT: *See also* the Snowshoe Hare account for a general description of rabbit scat.

Pygmy Rabbit   *Brachylagus idahoensis*

Pellets ¹/₈–⁹/₃₂ in. (0.3–0.7 cm) diameter

Cottontails and Brush Rabbits

Pellets ³/₁₆–⁷/₁₆ in. (0.5–1.1 cm) diameter

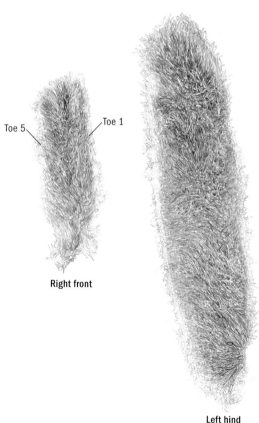

Toe 5

Toe 1

Right front

Left hind

Desert Cottontail feet.

Bounding pattern of
a Desert Cottontail
in which the front
tracks are placed
independently.

Bounding pattern of
a cottontail in which
the front feet are
placed together.

Sliding tracks of a Desert Cottontail slowing before committing to a stream crossing. The hind tracks fall to either side of the front tracks in the middle.

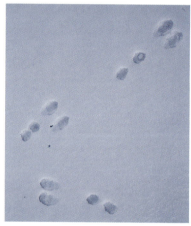

The classic bounding trails of Desert Cottontails in shallow snow.

ORDER SORICOMORPHA

Shrews: Family Soricidae

SHREWS | *Sorex* and *Notiosorex* spp.

Plate 1
Plate 13
Plate 88

TRACKS: Front: $3/16$–$5/16$ in. long (0.5–0.8 cm) × $3/16$–$5/16$ in. wide (0.5–0.8 cm)

Tiny. Slightly asymmetrical. Five toes. Toe 3 is longer than toes 2 and 4, and thus the tips of the three toes form an arch, or curve. The metacarpal region has six pads: four "palm" and two posterior "heel" pads. Not all pads register reliably. Claws register more often than not.

Hind: $^3/_{16}$–$^7/_{16}$ in. long (0.5–1.1 cm) × $^3/_{16}$–$^5/_{16}$ in. wide (0.5–0.8 cm)

Tiny. Symmetrical toes, asymmetrical metacarpal pads. Five toes. Toe 3 is equal in length to toes 2 and 4, and thus the tips of the three toes form a straight line. The metatarsal region has six pads: four "palm" and two posterior "heel" pads. Not all pads register reliably. Claws register more often than not. Tracks and trails of smaller shrew species are easily confused.

TRAIL:

Walk/trot:	Strides: 1–1½ in. (2.5–3.8 cm)	
	Trail widths: ¾–1 in. (1.9–2.5 cm)	
Straddle trot:	Strides: 1$^3/_{16}$–1$^3/_8$ in. (3–3.5 cm)	
	Trail widths: ¾–$^7/_8$ in. (1.9–2.2 cm)	
Bound:	Strides: 1–7 in. (2.5–17.8 cm)	
	Trail widths: $^{11}/_{16}$–1$^5/_{16}$ in. (1.7–3.3 cm)	
	Group lengths: $^5/_8$–2$^1/_8$ in. (1.6–5.4 cm)	

NOTES:

• The shrews of California leave similar sign that is difficult, if not impossible, to narrow to the species level.

• The numerous palm and heel pads in shrew tracks are a wonderful character to help in differentiating them from other small mammal species. Their tracks are also tiny, so small as to rule out most mice.

• In bounding trail patterns, the group length is usually very short (meaning the hind tracks barely overstep the front tracks), and there is often a gap between the front feet.

• Trot and straddle trot, covering ground quickly in their continuous hunting, only slowing to walk when foraging. They bound to cross open areas, when they feel exposed, and when traveling in deeper substrates, such as snow.

• Capable swimmers, and their sign is often found along riparian corridors.

SCAT: $^1/_{16}$–$^3/_{16}$ in. (0.2–0.5 cm) diameter, $^3/_{16}$–½ in. (0.5–1.2 cm) long

Shrew scats vary depending upon diet. When eating earthworms and slugs, shrew scats are soft, amorphous squirts. Scats containing harder substances are small and tubular, with tapered ends and some twisting. Look for scats along travel routes and runs, and accumulating at burrow entrances.

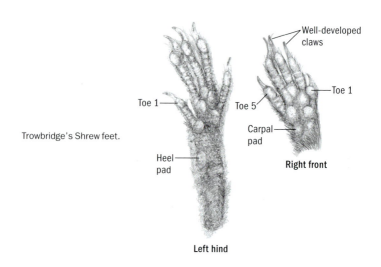

Trowbridge's Shrew feet.

Well-developed claws

Toe 1

Toe 5

Toe 1

Carpal pad

Heel pad

Right front

Left hind

Footprints of a shrew. The topmost and bottommost tracks are the front tracks, and the two in the middle the hind tracks.

The typical bounding trail of a shrew in light snow.

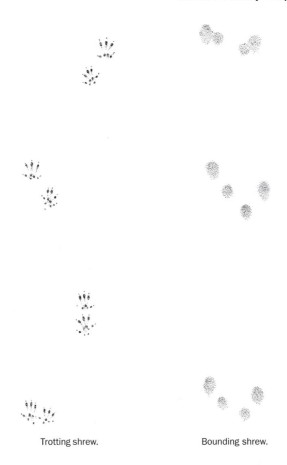

Trotting shrew.

Bounding shrew.

Moles: Family Talpidae

Plate 13
Plate 101

BROAD-FOOTED MOLE   *Scapanus latimanus*
TOWNSEND'S MOLE   *Scapanus townsendii*

TRACKS: Front: $^3/_8$–$^9/_{16}$ in. long (1–1.4 cm)

Tiny. Asymmetrical. Five toes. The entire foot is held in such a way that only toes 1, 2, and 3 register in tracks, and often just the claws.

Hind: $^3/_8$–$^1/_2$ in. long (1–1.3 cm) × $^1/_4$–$^7/_{16}$ in. wide (0.6–1.1 cm)

Tiny. Symmetrical. The metatarsal region may or may not register in tracks; all that is often visible is claws, and the digital pads.

TRAILS: Walk/wiggle trot: Strides: 1¼–1⅝ in. (3.2–4.1 cm)
Trail widths: 1½–2 in. (3.8–5.1 cm)

NOTES:

• When moles move above ground, they use a quick, writhing motion; it is difficult to decipher which gait is used. Walks and trots are probably used, in addition to the great deal of body undulations, which seem to help propel the animals forward.

• The trails of turtle hatchlings can look similar to mole trails, because the tiny front feet of turtles often register on their sides, leaving tracks similar to the front feet of moles. However, the hind track of moles is typical of small mammals and the most useful clue in distinguishing moles from turtles.

SCAT: ³/₃₂–³/₁₆ in. (0.2–0.5 cm) diameter, ³/₈–1 in. (1–2.5 cm) long

Mole scats are large, considering the size of the animal. Like those of shrews, mole scats vary depending upon diet. When eating earthworms and slugs, scats are soft squirts. Scats composed of harder substances are long, curved, and tubular with tapered ends and some twisting. It appears that certain mole species form above-ground latrines and others do not, but we do not definitively know which does which (Elbroch 2003). Latrines are found in the middle of fields but more often under cover, such as a log, exposed root, or even wooden stairs constructed along hiking trails.

URINE AND OTHER SCENT-MARKING BEHAVIORS: Anal gland secretions, and probably latrines as well, play a major role in olfactory communication of territorial boundaries (Hartman and Yates 2003). Urination almost always occurs within the burrows at tunnel junctions (Gorman and Stone 1990).

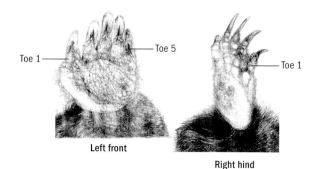

Broad-footed Mole feet.

Walk/trot of a mole.

A close look at the traveling trail of a Broad-footed Mole reveals the details of individual footprints.

A beautiful trail of the larger Townsend's Mole. Photo by George Leoniak.

Mole latrine.

ORDER CHIROPTERA: BATS

Free-tailed bats: Family Molossidae

Vesper bats: Family Vespertilionidae

SCATS:

Little Brown Myotis *Myotis lucifugus*

$^1/_{16}$–$^1/_8$ in. (0.2–0.3 cm) diameter, $^1/_8$–$^{11}/_{16}$ in. (0.3–1.7 cm) long

Pallid Bat *Antrozous pallidus*

$^1/_{16}$–$^5/_{32}$ in. (0.2–0.4 cm) diameter, $^1/_8$–$^{11}/_{16}$ in. (0.3–1.7 cm) long

Bat scats are dropped in flight and can be encountered anywhere. They are also dropped while roosting, and accumulations under regular and communal roosts can be massive. Beneath bridges with suitable cracks for roosting, and in caves, old mine shafts, attics, and abandoned buildings are common places to find scat accumulations. Bat scat, especially in mass, is referred to as *guano*.

Scats are typically tubular and twisted, but highly irregular, with rough surfaces and pointed ends. Shiny insect remains are typically easy to see in the scats of California's bats, most of which feed exclusively upon insects.

URINE AND OTHER SCENT-MARKING BEHAVIORS: Urine stains are common under bridges on concrete where bats cling while resting. Stains also form at the entrances to roosting sites and may play a role in chemical communication.

The latrine and midden of a Pallid Bat under a bridge in eastern Santa Barbara County, where the remains of large roaches lie among the accumulating feces.

Stains made by urinating *Myotis* spp. as they enter and exit a day roost in an old house.

ORDER CARNIVORA

Cats: Family Felidae

DOMESTIC CAT   *Felis catus*

Plate 7
Plate 19
Plate 109

TRACKS: Front: 1–1⅝ in. long (2.5–4.1 cm) × ⅞–1¾ in. wide (2.2–4.4 cm)

Small. Asymmetrical. Five toes: toe 1 is reduced and located above the metacarpal pads on the inner leg and is used for climbing and gripping prey. Four toes reliably register in tracks. The metacarpal pads are fused and create one large, trapezoidal pad; the leading edge is bilobed, and the posterior edge has three lobes. The impression made by the metacarpal pads is a very large proportion of the overall footprint. The negative space between the toes and metacarpals forms a "C." An additional carpal pad sits high on the leg and only registers at high speeds or in deep substrates. Claws are protractible (meaning when the muscles are relaxed, the claws are sheathed), thin, and decurved; claws do not often register in tracks. The front track is often larger and rounder than the hind track.

Hind: 1⅛–1½ in. long (2.9–3.8 cm) × ⅞–1⅝ in. wide (2.2–4.1 cm)

Small. Slightly asymmetrical. Five toes. The metatarsal pads are fused and create one large trapezoidal pad. The impression made by the metatarsal pads is proportionally smaller than in the front track, and is the key to differentiating front from hind tracks. The negative space between the toes and metatarsals is larger than that in front tracks, and forms a "C" or "H." The claws are protractible and do not often register. The hind track is more elongated and symmetrical than the front track, but can vary in size from similarly proportioned to much larger than the front track.

TRAIL: Walk: Strides: 6–12½ in. (15.2–31.8 cm)
Trail widths: 2–4¾ in. (5.1–12.1 cm)

Trot: Strides: 10–16 in. (25.4–40.6 cm)

Gallops: Strides: 14–40 in. (35.6–101.6 cm)
Group lengths: 10–33 in. (25.4–83.8 cm)

NOTES:

• Widespread, and can be found in national parks and other areas you might believe too remote. Often feral populations are subsidized by well-intentioned but misguided people who drop off food for them.

• Most common gait is an overstep walk. They direct-register walk in deep snow, creating trails similar to those made by Striped Skunks and Gray Foxes.

• Toes can appear more robust and blockier in tracks than those of Bobcats.

SCAT: ⅜–⅞ in. diameter (1–2.2 cm)

Cat scats are often long tubular ropes that may be segmented and folded in on themselves, or one great length. Cat scats generally have a smooth outer surface. Scats often have blunt ends, or just one pointy end, which was the last to exit the anus.

Cat scats are often found in scrapes and are often covered as well. House cats seem particularly attracted to soft substrates, "the sandbox phenomenon," in which it is easier to cover their scats. When house cats cover their scats, they pull material in toward the scat with their front paws, leaving scrapes of 7–10 in. (17.8–25.4 cm) in length.

URINE AND OTHER SCENT-MARKING BEHAVIORS: Various postures are employed by felids, including squatting and raising a leg, to deposit urine as scent posts. Domestic Cats also stand before an object of interest and squirt urine backward, like a squirt gun, with incredible accuracy. Look for trails that cut close to stumps, rocks, low-hanging foliage, lower branches, or a variety of other targets, and a rear foot to kick out to one side while they take aim and shoot. Domestic Cats usually spray urine between 8 and 9 in. (20.3–22.8 cm) off the ground.

As any house cat owner will attest, cats create scratching posts and use them time and again, shredding bark and sofas alike given enough time. Cats also have a gland on their cheeks that they rub against prominent objects and family members (Macdonald 1985).

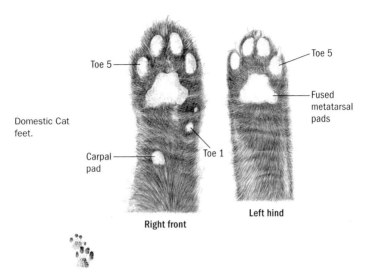

Domestic Cat feet.

Toe 5

Toe 5

Fused metatarsal pads

Carpal pad

Toe 1

Right front

Left hind

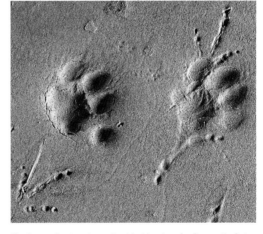

The larger front and smaller hind tracks of a Domestic Cat, along with accompanying crisscrossing American Robin trails.

Walking feral cat.

The carefully covered scat of a feral house cat under a bridge.

The scratch post of a Domestic Cat.

BOBCAT

Lynx rufus

Plate 11
Plate 21
Plate 113

TRACKS: Front: 1⁵/₈–2³/₈ in. long (4.1–6 cm) × 1³/₈–2½ in. wide (3.5–6.4 cm). Tracks can splay to 3 in. (7.6 cm) wide.

Small to medium. Asymmetrical. Five toes. Toe 1 is reduced and located above the metacarpal pads on the inner leg and is used for climbing and gripping prey. Four toes reliably register in tracks. The metacarpal pads are fused and create one large trapezoidal pad; the leading edge is bilobed, and the posterior edge has three lobes. The impression made by the metacarpal pad is a very large proportion of the overall footprint. The negative space between the toes and metacarpals forms a "C." An additional carpal pad sits high on the leg, and only registers at high speeds or in deep substrates. Claws are protractible, thin, and decurved; claws do not often register in tracks. The front track is larger and rounder than the hind track.

Hind: 1⁹/₁₆–2³/₈ long (4–6 cm) × 1³/₁₆–2½ in. wide (3–6.4 cm)

Small to medium. Slightly asymmetrical. Four toes. The metatarsal pads are fused and create one large trapezoidal pad. The impression made by the metatarsal pad is proportionally smaller than in the front track. The negative space between the toes and metatarsals is larger than that in front tracks, and forms a "C" or "H." The claws are protractible and do not often register. The hind track is more elongated and symmetrical than the front track.

TRAIL:

Direct-register walk:	Strides: 6–14 in. (15.2–35.6 cm)	
	Trail widths: 5–9½ in. (12.7–24.1 cm)	
Overstep walk:	Strides: 12–18 in. (30.5–45.7 cm)	
	Trail widths: 3¾–5 in. (9.5–12.7 cm)	
Trot:	Strides: 15–26 in. (38.1–66 cm)	
	Trail widths: 3–4½ in. (7.6–11.4 cm)	
Gallop:	Strides: 16–48 in. (40.6–121.9 cm)	
	Group lengths: 35–45 in. (88.9–114.3 cm)	

NOTES:

- Walk when hunting and traveling, only speeding up to trots, bounds, and gallops when they feel exposed, are at risk of being hunted, or are chasing prey. Like Cougars, they meander more when foraging or looking for potential prey, and when traveling, move in straighter lines.

- Often move within cover, or travel along the edge of open habitats.

- Direct-register walk in deep snow, and like Domestic Cats, leave a beautiful pattern of triangles created by the sweeping of the legs as they move.

- The positions of toes 3 and 4 (the inner toes) on the hind feet are useful in differentiating the tracks of Bobcats and Domestic Cats. In Bobcats, toes 3 and 4 are farther forward than toes 2 and 5, while in Domestic Cats toes 3 and 4 are positioned closer to the metatarsal pad and more in line with toes 2 and 5.

DIFFERENTIATING COUGAR AND BOBCAT TRACKS

When in doubt as to whether you have found the tracks of a very large Bobcat or a very young Mountain Lion, use the width of the metacarpal-metatarsal pads. The cutoff is 1⅝ in. (4.1 cm). Anything larger is almost certainly a Cougar and anything smaller a Bobcat or Domestic Cat.

SCAT: ⁷/₁₆–1 in. (1.1–2.5 cm) diameter, 3–9 in. (7.6–22.9 cm) long

Bobcat scats are often long tubular ropes that may be segmented and folded in on themselves, or one great length. Bobcat scats have a deceptively smooth outer surface with a twisted interior you only see when you break them open; scats of fur and bone are often difficult to break apart, because the contents are so compacted and interlaced in the interior. Unlike Coyote scats, they are difficult to squash unless very fresh. Bobcat scats are often segmented in appearance and have blunt ends, or just one pointy end, which was the last to exit the anus. Bobcats are strictly carnivorous; however, grasses are consumed at times and may appear in scats.

As opposed to Cougar scat, Bobcat scat is quite easy to find on the landscape. At high densities (as they are in many parts of California), Bobcats form conspicuous latrines along travel routes, especially at game trail junctions. Bobcat latrines may have over 50 individual scats in an area of 1 m diameter (Bailey 1974). In a study of 60 scats found while trailing Bobcats in snow in Idaho, 60 percent were covered and 40 percent were exposed (Bailey 1974). When a Bobcat covers a scat, it does so by pulling material in with its front paws, leaving scrapes 12–18 in. (30.5–45.7 cm) in length. In our experience in California, they cover their scats much less often than this. Scats are often placed in scrapes, or they may be left without adornment.

URINE AND OTHER SCENT-MARKING BEHAVIORS: Various postures are employed by felids to deposit urine as scent posts, including squatting and raising a leg. Bobcats in particular most often stand before an object of interest, and spray urine backward, like a squirt gun, with incredible accuracy. Look for trails that cut close to stumps, rocks, low-hanging foliage, lower branches, or a variety of other targets, and a rear foot to kick out to one side while they take aim and shoot. Bobcats usually spray urine between 9 and 17 in. (22.9–43.2 cm) off the ground.

In dry habitats, accumulated urine sprays at regular scent posts on rocks or tree trunks can be identified by the dark stains that form over time. On three occasions while trailing Bobcats in snow, tracker John McCarter documented 32 urine scent posts and a scat in less than a mile (1.6 km) of trail, 23 urine scent posts and a scat in half a mile (0.8 km), and on one incredible occasion, 31 urine scent posts and two scats in less than a quarter of a mile (0.4 km). The gender and breeding status of these Bobcats were unknown.

Like Cougars, Bobcats create small scrapes and mounds with their hind feet while the cat is stationary; first a cat makes a controlled scrape with one hind foot, quickly followed by the other, and the process is repeated several times, leaving two parallel marks divided by only the thinnest of ridges and a neat pile of debris at one end. This is a scent-marking behavior in and of itself, but a scrape may also be accompanied by urine or scat. This is the behavior most biologists and trackers are referring to when they employ the word *scrape*.

Debris mounds are often found under rock overhangs, along runs that follow vertical rock ledges, or adjacent to well-used trails in suitable leaf litter. These mounds may also be placed next to a scent post made by spraying urine backward at some target. The scrapes are 3–7½ in. (7.6–19.1 cm) wide. Bailey (1974) found that while adults scent-mark with feces, urine, scrapes, and anal glands, young Bobcats did not make scrapes or post other chemical communication. Kittens always attempted to cover their feces as well.

Like Domestic Cats and Cougars, Bobcats occasionally make scratch posts, leaving scent from the glands on their feet. However, they appear to make them very infrequently, and they are very difficult to find. Felids also have a gland on their cheeks that they rub against prominent objects and family members (Macdonald 1985).

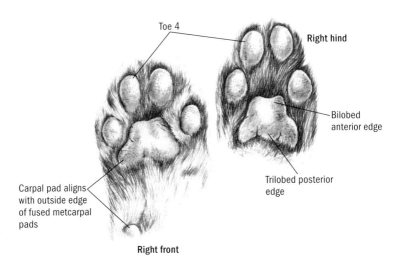

Toe 4

Right hind

Bilobed anterior edge

Carpal pad aligns with outside edge of fused metcarpal pads

Trilobed posterior edge

Right front

Bobcat feet.

Overstep walk
of a Bobcat.

Direct-register
walk of a Bobcat.

Trotting
Bobcat.

Loping Bobcat.

Front and hind tracks of a Bobcat, and above them, the front track of a Coyote.

The distinctive round circles and wide trail width characteristic of a direct-registering Bobcat.

A fresh Bobcat scrape and scat in moist sand along a road in Seaside, California.

In this instance, a Bobcat scraped with both the front and hind feet. First the animal scraped with its hind feet (look for the footprint). Second, the cat defecated in the depression. Last, the Bobcat scraped toward the scat with its front feet on three sides of the scat, yet never covered it with any dirt.

A Bobcat pauses to mark an old stump by spraying urine backwards onto it, and then bounds off up the hill.

COUGAR (MOUNTAIN LION) *Puma concolor*

Plate 12
Plate 30
Plates 115 and 123

TRACKS: Front: 2¾–3⅞ in. long (7–9.8 cm) × 2⅞–4⅞ in. wide (7.3–12.4 cm)

Medium to large. Asymmetrical. Five toes. Toe 1 is reduced (but the claw is massive) and located above the metacarpal pads on the inner leg; it is used for climbing and gripping prey. Four toes reliably register in tracks. The metacarpal pads are fused and create one large trapezoidal pad; the leading edge is bilobate, and the posterior edge has three lobes. The impression made by the metacarpal pads is a very large proportion of the overall footprint. The negative space between the toes and metacarpals forms a "C." An additional carpal pad sits high on the leg and only registers at high speeds or in deep substrates; the carpal pads of Cougars are especially large and well developed to aid in gripping prey. Claws are protractible, thin, and decurved; claws do not often register in tracks. The front track is larger and rounder than the hind one.

Hind: 3–4⅛ in. long (7.6–10.5 cm) × 2⁹⁄₁₆–4⅞ in. wide (6.5–12.4 cm)

Medium to large. Slightly asymmetrical. Four toes. The metatarsal pads are fused and create one large trapezoidal pad. The impression made by the metatarsal pad is proportionally smaller than in the front track. The negative space between the toes and metatarsals is larger than that in front tracks and forms a "C" or "H." The claws are protractible and do not often register. The hind track is more elongated and symmetrical than the front track.

TRAIL:	Overstep walk:	Strides: 19–32 in. (48.3–81.3 cm)
		Trail widths: 5–9 in. (12.7–22.9 cm)
	Direct-register walk:	Strides: 15–28 in. (38.1–71.1 cm)
		Trail widths: 4–11 in. (10.2–27.9 cm)
	Trot:	Strides: 29–38 in. (73.7–96.5 cm)
		Trail widths: 3–5½ in. (7.6–14 cm)
	3× lope:	Strides: 45–55 in. (114.3–139.7 cm)
		Group length: 40–50 in. (101.6–127 cm)
	Gallop:	Strides: 36–120 in. (91.4–304.8 cm), occasionally modified bounds of up to 25 ft (762 cm)
		Group length: 50–75 in. (127–190.5 cm)

NOTES:

• Walk most of the time, twisting and turning when hunting, and moving in straighter lines when traveling. They often use an overstep walk in shallow substrates and a direct-register walk in deeper substrates.

• Frequently use dirt roads and trails to travel their range, as well as dry washes and other avenues of least resistance.

- When there are no claws showing, Domestic Dog tracks are often mistaken for Lion tracks. Look at the relative size and shape of the metacarpal vs. metatarsal pads. In dogs the palm pad is about two to three times larger than one of the toes, while in Lions it is closer to four times larger. Also note whether the front edge of the palm pad is pointed (canid characteristic) or two-pointed, or bilobate (felid characteristic).

- Distinguishing male from female Cougars is discussed in Chapter 3, in the section "Determining the Gender of an Animal."

SCAT: ¾–1⅝ in. (1.9–4.1 cm) diameter, 6½–17 in. (16.5–43.2 cm) long

Cougar scats are often tubular, long ropes that may be segmented and folded in on themselves, or one great length. Cougar scats sometimes loosely twist like those of Wolves, but in general, cat scats have a deceptively smooth outer surface with a twisted interior you only see when you break them open. Tubular Cougar scats may have blunt ends or one pointy end, which was the last to exit the anus. Cougars are strictly carnivorous; however, grasses are consumed at times and appear in scats.

Cougar scats vary depending upon the size of their prey, and if they are eating a large mammal, how much of the carcass they have consumed. The first scats after opening a large carcass and feeding on the internal organs are typically soft and amorphous. As they transition from internal organs to meat, scats are tubular but soft, and they crumble easily when they dry out and turn white. For smaller prey and in the last stages of large-carcass consumption, scats are tubular and sometimes tapered.

Unlike Bobcat scats, Cougar scats can be surprisingly difficult to find. This may be due to the feast-and-famine feeding rituals experienced by many Cougars, and that food moves through their system relatively quickly. Scats are most easily found adjacent to kill sites in nearby latrines under cover (like the center of thick bushes), or en route between a kill site and bedding area. Scats are also placed along travel routes as they patrol their territories, sometimes in scrapes, occasionally covered but more often left as is. When a Cougar covers its scats, it uses its front feet to pull in material around a scat, leaving scrapes of 20–36 in. (50.8–91.4 cm) in length.

URINE AND OTHER SCENT-MARKING BEHAVIORS: Cougars often squat to deposit urine, and as far as we know, do not spray urine backwards on objects as do Domestic Cats, Bobcats, and Jaguars.

Like Bobcats, Cougars create scrapes with the *hind feet* while they are stationary. First the Cougar scrapes with one hind foot and then the other, and the process is repeated multiple times to create a neat pile of debris and two parallel marks divided by only the thinnest of ridges. This is a scent-marking behavior in and of itself, but a scrape may also be accompanied by urine or scat. This is the behavior most biologists and trackers are referring to when they employ the word *scrape*.

During intensive tracking of marked individuals in California and Chile, Mark Elbroch has noted that while all male Cougars make scrapes with regularity, certain females mark more than others. He noted that some females both with and without kittens made scrapes frequently, while others were never documented making scrapes during the studies (Elbroch, unpublished data). Maehr (1997) hypothesizes that females may only make scrapes while they are in estrus. In a study by Seidensticker et al. (1973) about 20% of scrapes were accompanied by a scat or urine; however, they acknowledge that this may be an underestimate because of the difficulty of detecting urine. Harmsen et al. (2010) concluded that scrapes are often made in response to discovering the scrape made by another Cougar, but that scrapes were not created to convey dominance. Scrapes measure 5½–8 in. (14–20.3 cm) in width for females and 7–11 in. (17.8–27.9 cm) across for males. The deeper the substrate, the wider the scrape.

Cougars also scratch trees and logs as part of chemical communication. In following known marked Cougars, Elbroch (unpublished data) has only recorded this behavior for male Cougars, and only six times in several years of intensive tracking. Felids also have a gland on their cheeks that they rub against prominent objects and family members (Macdonald 1985).

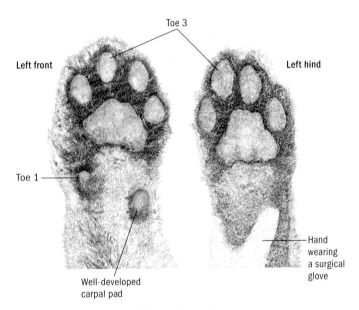

Adult male Cougar feet.

Bounding Cougar.

Overstep walk of a Cougar.

Light front right track on the left, and hind right track on the right of an adult female Cougar in a film of mud.

Left front and hind tracks of an adult male Cougar in moist snow, which has since frozen and set.

Multiple Cougar scrapes at a regular marking area beneath a Ponderosa pine.

Look closely for the scratches of a male Cougar marking its territory on a burnt palm log in Anza Borrego State Park in eastern San Diego County.

Dogs: Family Canidae

DOMESTIC DOG   *Canis lupus familiaris*

Plate 114

TRACKS: Front: All sizes.

Small to large. Digitigrade. Symmetrical. Five toes. Toe 1 is greatly reduced, is located above the metacarpal pads on the inner leg, and generally only registers when the animal is running. Toes 2–5 reliably register in tracks. The metacarpal pads are fused and create one triangular pad. The negative space between toes and metacarpals forms an "H" more often than an "X." An additional carpal pad sits on the same plane as toe 1, and only registers at high speeds or in deep substrates. Claws register reliably, and are heavy and sharp. Front larger and rounder than hind track.

Hind: All sizes.

Small to large. Digitigrade. Symmetrical. Four to five toes. Toe 1, present only in some individuals, is greatly reduced and located above the metatarsal pads on the inner leg, and generally does not register. The metatarsal pads are fused and create one triangular pad. The negative space between the toes and metatarsals forms an "X." Claws register reliably. The hind track is smaller, narrower, and more elongated than the front track.

NOTES:

- Highly variable, can range in size from smaller than a Gray Fox to larger than a Cougar.

DIFFERENTIATING DOG AND COYOTE TRACKS

Differentiating the tracks of dogs and Coyotes can sometimes be very difficult, so use the following guidelines:

Toe Shape and Position The toes of a Coyote track are narrow and somewhat teardrop-shaped in comparison to those of dogs, which are usually more rounded or blocky. Coyote toes also tend to point straight forward, while those of dogs tend to splay outward. The outer two toes (toes 2 and 5) in Coyote tracks are usually positioned tight into the track and are pointed forward. In the hind feet, the outer toes may even be behind the inner toes. In dogs, toes 2 and 5 are set further out and angle outward, contributing to the rounder appearance of Domestic Dog tracks.

Claws Coyote claws tend to be small and sharp, while dog claws are comparatively large and blunt. It is common for the claw marks from the outer toes (toes 2 and 5) in Coyote tracks not to register or to register so close to toes 3 and 4 that they are easily overlooked.

Metacarpal-metatarsal pads Coyotes lean forward in their tracks, so the toes register comparatively deeper than in dogs. Because of this tendency, the palm pads register more lightly and they appear smaller than those of dogs. This is even more pronounced in the hind feet of Coyotes, where the metatarsals may register as a mere dot the same size as one of the toes. In deeper substrates, two small lobes are visible on either side of the central pad (sometimes called "wings") in the hind feet of Coyotes.

As a general rule (one that can occasionally be broken), the metacarpal pad of the Coyote is as long or longer than it is wide, and in most Domestic Dog tracks the metacarpal pad is wider than long. This can vary due to substrate, so use common sense and compare several front tracks to be sure.

Trail characteristics Coyote trails are "cleaner." They direct-register far more often than dogs and leave narrower trail widths. Their trails are also straighter, and they waste less energy than dogs on a typical day in the woods.

SCAT: All sizes

Just like the dogs themselves, their scats come in every size. They are typically tubular, and twisted, and have tapered ends. Like all animals, their scats are reflective of their diet. Scats composed of processed dog foods are typically soft tubes that fold in and back on themselves. Not to be gross, but large dog scats can resemble human scats as well, because they often share in our diet. When dogs eat wild foods, their scats resemble those of foxes, Coyotes, and Wolves.

Dogs often defecate in areas where scent-marking is less important—especially when contained in houses for much of the day and relegated to specific areas and times when they can defecate.

URINE AND OTHER SCENT-MARKING BEHAVIORS: Like other canids, dogs use urine to mark prominent objects as they travel, including stumps, fire hydrants, and car tires.

Like Coyotes, Domestic Dogs have scent glands between their toes and they often scratch adjacent to scats or urine. Scratches are typically narrow, short scrapes in the earth, leaf litter, or snow, and they may throw debris up to 10 ft (3 m) behind them. Unlike the controlled scraping of cats, canids make scrapes with rapid, repetitive, and forceful swipes of their paws. These scrapes are more often done with only the back feet, but may also be with both front and hind paws, leaving two depressions in the ground. Dog scrapes are common along hiking trails where they are frequently walked by their owners.

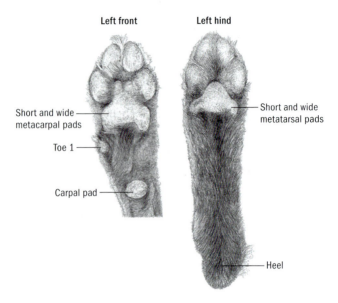

Left front   Left hind

Short and wide
metacarpal pads

Toe 1

Carpal pad

Short and wide
metatarsal pads

Heel

Domestic Dog feet.

Front and hind tracks of a Domestic Dog.

COYOTE   *Canis latrans*

Plate 28

Plates 114 and 122

TRACKS: Front: 2¼–3¼ in. long (5.7–8.3 cm) × 1½–2½ in. wide (3.8–6.4 cm)

Medium. Digitigrade. Symmetrical. Five toes. Toe 1 is greatly
reduced, is located above the metacarpal pads on the inner leg,
and generally only registers when the animal is running. Toes 2–5

reliably register in tracks. The metacarpal pads are fused and create one triangular pad; the posterior edge may include "wings," formed by two lobes that project backwards. The negative space between the toes and metacarpals forms an "X." An additional carpal pad sits on the same plane as toe 1 and only registers at high speeds or in deep substrates. Claws register reliably, although the claws on toes 2 and 5 register less so and are often so close to toes 3 and 4 that they are overlooked. Coyote claws are very fine and thin, and those on toes 3 and 4 often point inward toward each other and register very close together. Front track is larger than hind track.

Hind: 2⅛–3 in. long (5.4–7.6 cm) × 1⅛–2 in. wide (2.9–5.1 cm)

Medium. Digitigrade. Symmetrical. Four toes. The metatarsal pads are fused and create one triangular pad; the metatarsal pads often register incompletely, appearing as a single circle no larger than one of the digital pads. When the metatarsal pads register completely, the central circular pad is sandwiched between two additional, very small lobes on either side. The negative space between the toes and metatarsals forms an "X." Claws register reliably, although the claws on toes 2 and 5 register less so, and often so close to toes 3 and 4 that they are overlooked. All claws are very fine and thin. Claws on toes 3 and 4 often point inward toward each other and register very close together.

TRAIL:	Trot:	Strides: 15–21 in. (38.1–53.3 cm)
		Trail widths: 2¼–4 in. (5.7–10.2 cm)
	Walk:	Strides: 11–17 in. (27.9–43.2 cm)
		Trail widths: 4–5½ in. (10.2–14 cm)
	Side trot:	Strides: 19–27 in. (48.3–68.6 cm)
		Group lengths: 4½–11½ in. (11.4–29.2 cm)
	Lope:	Strides: 5–20 in. (12.7–50.8 cm)
		Group lengths: 26–33 in. (66–83.8 cm)
	Gallop:	Strides: 12–90 in. (30.5–228.6 cm)
		Group lengths: 35–101 in. (88.9–256.5 cm)

NOTES:

- Splayed tracks in soft substrates are more difficult to differentiate from Domestic Dogs.

- Trot when traveling or exploring, and slow down into a walk when stalking small prey or investigating a smell or sound.

- Use a direct-register walk in deep snow, and unlike in felids, the drag lines between tracks are straight.

- Social animals, and although single transients are natural in every population, you will find pairs and packs traveling together throughout the year.

- Advice for distinguishing between Domestic Dog and Coyote tracks is found under the Domestic Dog account.

SCAT: $^3/_8$–$1^3/_8$ in. (1–3.5 cm) diameter

Coyote scats vary tremendously, depending upon diet. Scats of fruit and mast crops are tubular and typically of a larger diameter and volume than meat scats; fruit scats exhibit little to no twisting and a smooth surface, and they have either blunt or tapered ends. Scats composed completely of meat and internal organs can be completely amorphous or can form loose tubes of crumbly material that break easily. Scats of hair and bone, which last the longest in the wild, are twisted and tapered ropes with pointy ends. Any bone shards that Coyotes consume must be accompanied by hide and hair to cushion and protect the innards from potential punctures.

Coyotes post scats in obvious places, often elevated, and in the middle of trails or roads. Trail junctions are heavily marked as well, where they may accumulate given time. Scats may also accumulate along regular travel routes, and near regular bedding sites and dens.

Generally, scats greater than ¾ in. (1.9 cm) in diameter rule out Red Foxes, but lacking contextual clues or associated signs, Coyote and fox scats can be difficult to differentiate. Scats are deposited equally throughout a Coyote's range, whereas urinations and scratches are made most often along the periphery of their territories (Gese and Ruff 1997).

URINE AND OTHER SCENT-MARKING BEHAVIORS: Coyotes urinate on elevated surfaces, old stumps, low foliage, and anywhere else that might hold and spread scent. They use urine to scent-mark territories and communicate sexual status and pack status, with one another.

When the female is in estrus, her urine is colored orange to red by blood. When in full estrus, blood may drip from the vulva as she moves. Author and tracker Paul Rezendes shared pictures with me of a female Coyote's bed, where puddles of blood had formed when she lay down.

During estrus, pair bonding becomes especially apparent; look for double scent posts, one with blood in the urine and the other without. Often one animal squats and the other raises a leg. A study by Gese and Ruff (1997) found that 60 percent of urinations by alpha males were with a raised leg and 25 percent while in a standing, "forward-lean" position, both typical dominance displays. Ninety-two percent of female urinations were done while squatting. They also found that the females in alpha pairs initiate double-marking 75 percent of the time.

Coyotes have scent glands between their toes, and they often scratch adjacent to scats or urine. Scratches are narrow, short scrapes in earth or snow, where debris may be thrown up to 10 ft (3 m). Unlike the controlled scraping of cats, canids make scrapes with rapid, repetitive, and forceful swipes of their paws. These scrapes are more often done with only the back feet but may also be with both front and hind paws, leaving two depressions in the ground. Transient Coyotes (those without territories) scent-mark much less frequently than residents, and in one study were never observed to scratch the ground after defecations or urinations, a behavior frequently exhibited by Coyotes holding a territory (Gese and Ruff 1997).

Alpha pairs are the primary defenders of a pack's territory and scent-mark much more frequently along the edge of their territories than in the interiors. They repeatedly scent-mark areas prone to intrusion by neighboring Coyotes (Gese and Ruff 1997).

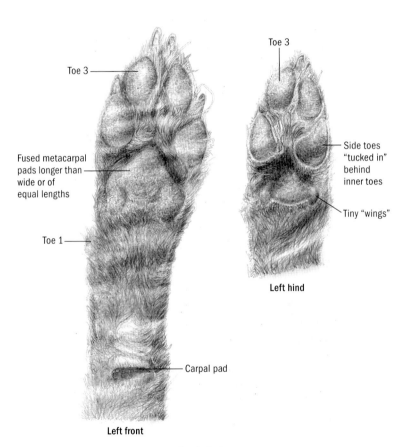

Coyote feet.

Galloping
Coyote.

Overstep walk
of a Coyote.

Direct-register
trot of a
Coyote.

Side trot of a
Coyote.

The front track of this Coyote sits behind the hind track. Note the "wings" on the palm pad of the hind track.

Perfect front (ahead) and hind tracks of a Coyote in moist sand. Note the front track's characteristic asymmetry.

Coyote tandem marking while the female is in estrus; her urine is the color of blood. Also note the accompanying scrapes.

Scratches made by a Coyote after marking a trail junction.

WOLF   *Canis lupus*

Plate 33

Plate 116

TRACKS: Front: 3¾–5¾ in. long (9.5–14.6 cm) × 2⅞–5 in. wide (7.3–12.7 cm)
There are reports of 6-in. (15.2-cm) tracks.

Large. Digitigrade. Symmetrical. Five toes. Toe 1 is greatly reduced
and located above the metacarpal pads on the inner leg, and generally only registers when the animal is running. Toes 2–5 reliably
register in tracks. The metacarpal pads are fused and create one triangular pad. The negative space between toes and metacarpals forms
an "H" more often than an "X." An additional carpal pad sits on the
same plane as toe 1 and only registers at high speeds or in deep substrates. Claws register reliably, and are heavy and sharp. Front tracks
are larger and rounder than hind.

Hind: 3¾–5¼ in. long (9.5–13.3 cm) × 2⅝–4½ in. wide
(6.7–11.4 cm)

Large. Digitigrade. Symmetrical. Four toes. The metatarsal pads are
fused and create one triangular pad. The negative space between the
toes and metatarsals forms an "X." Claws register reliably. The hind
track is smaller, narrower, and more elongated than the front track.

TRAIL: Direct-register trot: Strides: 22–34 in. (55.9–86.4 cm)
                                     Trail widths: 4–9¼ in. (10.2–23.5 cm)

           Side trot: Strides: 26–39 in. (66–99.1 cm)
                                     Group lengths: 7–15½ in. (17.8–39.4 cm)

           Walk: Strides: 13–24 in. (33–61 cm)
                                     Trail widths: 6–10 in. (15.2–25.4 cm)

Lope:	Strides: 20–23 in. (50.8–58.4 cm)
	Group lengths: 40–49 in. (101.6–124.5 cm)
Gallop:	Strides: 6–68 in. (15.2–172.7 cm)
	Group lengths: 55–99 in. (139.7–251.5 cm)

NOTES:

• Recolonizing old range; confirmed accounts of Wolves now come from as far south as Oregon. It is likely just a matter of time before confirmed cases appear in northern California.

• Trot incredible distances while exploring and patrolling home ranges. Like Coyotes, they use numerous gaits depending upon mood and behaviors.

• Trot may be duck-toed, with tracks pointing outward.

• Use a direct-register walk in deep snow (although they have the strength to continue to trot to impressive depths).

• Social animals, and although single transients are natural in every population, you will find pairs and packs traveling together throughout the year. Pack size varies from four animals to over 30, depending on available resources, population density, and numerous other variables.

SCAT: ½–1⁷/₈ in. (1.3–4.8 cm) diameter, 6–17 in. (15.2–43.2 cm) long

Wolf scats are tubular, and have either blunt or tapered ends. Scats composed completely of meat and internal organs can be completely amorphous, or loose tubes of crumbly material that break easily. Scats of hair and bone, which last the longest in the wild, are twisted and tapered ropes with pointy ends. Any bone shards that Wolves consume must be accompanied by hide and hair to provide protective cushioning to the innards from potential punctures.

Wolves post their scats on elevated surfaces, or in the middle of trails or roads. Trail junctions are heavily marked as well, where they may accumulate over a period of time. Scats may also accumulate along regular travel routes, and near regular bedding sites, rendezvous sites, and dens. Captive Wolves often defecate most where their caretakers enter their enclosures, as if to create an olfactory fence to inhibit intrusion (Harrington and Asa 2003). Anal sac secretions may or may not be deposited on feces. Little is known about their function, yet alpha males deposit anal secretions more than other Wolves, and most often before and during the breeding season (Asa et al. 1985).

URINE AND OTHER SCENT-MARKING BEHAVIORS: Wolves urinate on any elevated surfaces that might hold and spread scent, including old stumps, low foliage, and boulders. They use urine to mark territories and communicate sexual status and pack status. Scent-marking is placed approximately every 787 ft (240 m) throughout their range, though with twice the frequency along territory borders as within core areas. In addition to territorial functions, urine marking probably conveys information on reproductive and sexual status, as well as individual identity (Harrington and Asa 2003). Alpha males mark the most frequently, and raise their hind legs high when urinating. The alpha female too often raises her hind leg, although the stance is somewhat different. She pulls her leg up and forward more than straight up. Subordinate and juvenile females squat in typical canine fashion, and juvenile and subordinate males stretch forward, straightening both their hind legs in their version of a squat.

The alpha pair scent-mark together, called double or tandem marking, more frequently before and during the breeding season. Newly formed pairs tandem mark even more than pairs that have been together for several years. When the alpha female is in estrus, her urine is tinged by blood with orange or red.

Wolves sometimes stand stiff-legged, and scratch and rake the earth and debris adjacent to urine or scat. This could leave additional scents from glands found between the toes, as well as smells from the fresh earth.

The trail of a Wolf in an overstep walk—the tracks from back to front in the direction the animal is moving is front, hind, front, hind. Note the Coyote and Canada Goose tracks crossing at the center.

–5 relia—
eate one
the toes and
the same
bstrates.
tracks

Overstep walk
of a Wolf.

Side trot of a
Wolf.

Loping track
pattern of a Wolf.

Trotting Wolf.

GRAY FOX
ISLAND FOX

Plate 8
Plate 19
Plate 108

TRACKS: Front: 1³/₈–1¾ in. long (3.5–4.4 cm) × 1³/₁₆–1⁵/...

Small. Digitigrade. Symmetrical. Five toes. Toe 1...
and located above the metacarpal pads on the inne...
ally only registers when the animal is running. Toes...
register in tracks. The metacarpal pads are fused and c...
rounded, yet triangular pad. The negative space betwee...
metacarpals forms an "H." An additional carpal pad sits o...
plane as toe 1 and only registers at high speeds or in deep st...
Claws are semiprotractible and may or may not register. Fron...
are larger and rounder than hind ones.

Hind: 1¼–1⁵/₈ in. long (3.2–4.1 cm) × ¹⁵/₁₆–1³/₈ in. wide (2.4–3.5 c...

Small. Digitigrade. Symmetrical. Four toes. The metatarsal pads are
fused and create one rounded, yet triangular pad; the metatarsal pads
often only partially register. The negative space between the toes and
metatarsals forms an "X." Claws are semiprotractible and may or
may not register. The hind track is smaller and more elongated than
the front track.

TRAIL:

Walk:	Strides: 6–10 in. (15.2–25.4 cm)	
	Trail widths: 3–4½ in. (7.6–11.4 cm)	
Trot:	Strides: 9–18 in. (22.9–45.7 cm)	
	Trail widths: 1⁷/₈–4 in. (4.8–10.2 cm)	
Straddle trot:	Strides: 13–16 in. (33–40.6 cm)	
	Group lengths: 3½–4½ in. (8.9–11.4 cm)	
Lope:	Strides: 15–22 in. (38.1–55.9 cm)	
	Group lengths: 16–20 in. (40.6–50.8 cm)	
Gallop:	Strides: 13–68 in. (33–172.7 cm)	
	Group lengths: 15–51 in. (38.1–129.5 cm)	

NOTES:

- Move about their range in a trot, but will also travel roads and cross open areas in a straddle trot. In tracks made by trotting Gray Foxes, it is typical for the toes of the front tracks to be visible in front of the complete hind tracks.

- Capable climbers, although they generally climb angled trunks and saplings, as well as those trees with ample lower limbs. They sometimes bed in trees and shrubs. Their trails are often found along the tops of logs, walls, fences, and other raised surfaces; Gray Foxes have a catlike agility.

- Often use holes, or slash piles or other cover for bedding purposes and rarely lie out in the open.

- Struggle in deep snow, and hole up for long periods when conditions are soft, awaiting sun and wind that compacts and firms the snow surface to better support them. When snowmobile or cross-country trails are available they will use these as their primary travel routes, taking short excursions off to either side to hunt and forage.

- Most frequently confused with Domestic Cat tracks. The greater symmetry and proportionally smaller metacarpal/metatarsal pads help differentiate foxes from cats.

SCAT: $^{5}/_{16}$–¾ in. (0.8–1.9 cm) diameter, 3–6 in. (7.6–15.2 cm) long

Neither color nor size alone provides the necessary information to distinguish Gray from Red fox scats reliably; both species eat fruits, mast, and meat. Fox scats vary tremendously, depending upon diet. Scats of fruit and mast crops are tubular and typically of a larger diameter and volume than meat scats; fruit scats exhibit little to no twisting, a smooth surface, and either have blunt or tapered ends. Scats of hair and bone, which last the longest in the wild, are twisted and tapered ropes with pointy ends.

Like other canids, foxes post scats along trails and roads, often on elevated surfaces. Gray Foxes create latrines. A study in Utah found that 17 percent of Gray Fox scats occurred in latrines of two to eight scats, and 72 percent were deposited singly (Trapp 1978).

URINE AND OTHER SCENT-MARKING BEHAVIORS: The smell of Gray Fox urine is less pungent than that of Red Foxes; however, when it accumulates at regular scent posts it can be strong. Like other canids, foxes urinate on any elevated surface that might hold and convey scent.

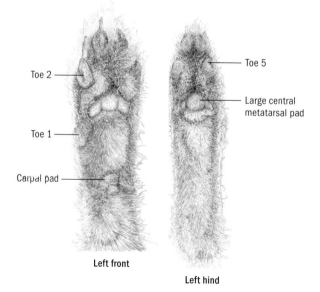

Common Gray Fox feet.

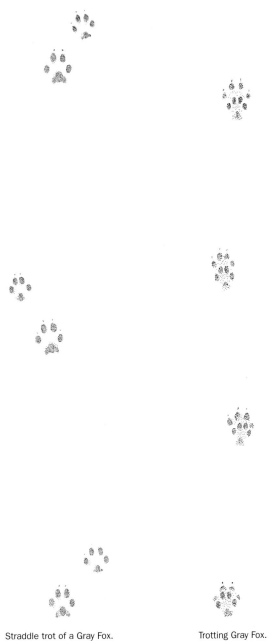

Straddle trot of a Gray Fox.

Trotting Gray Fox.

The toes and claws of the front track of this Gray Fox are visible, but the marks made by the metacarpal pads are completely covered by the smaller hind track that registers just behind.

A wonderful comparison between the similar tracks of a Gray Fox (in the middle) and a house cat (tracks both before and after the fox).

A latrine made by a family of Gray Foxes adjacent to an active den north of Santa Cruz, California.

KIT FOX   *Vulpes macrotis*

Plate 18
Plate 105

TRACKS: Front: 1–1$^{11}/_{16}$ in. long (2.5–4.3 cm) × $^{15}/_{16}$–1½ in. wide (2.4–3.8 cm)

Small. Digitigrade. Symmetrical. Five toes. Toe 1 is greatly reduced and raised up on the inner leg. Four toes reliably register in tracks. The metacarpal pads are fused and create one triangular pad. The marks from the digital and metacarpal pads are sometimes obscured by additional fur growing on the feet; some Kit Foxes have furred soles, while others do not. The negative space between toes and metacarpals forms an "H." There is an additional carpal pad higher on the leg. Claws may or may not register. The front track is larger and rounder than the hind.

Hind: 1⅛–1⅝ in. long (2.9–4.1 cm) × ⅞–1¼ in. wide (2.2–3.2 cm)

Small. Digitigrade. Symmetrical. Four toes. The metatarsal pads are fused and create one triangular pad; the metatarsal pads may only partially register. The marks from the digit and metatarsal pads may or may not be obscured by additional fur growing on the soles; some Kit and Swift foxes (*V. velox*) have furred feet, while others do not. The negative space between toes and metacarpals forms an "X." Claws may or may not register. The hind track is more elongated than the front track.

TRAIL:

Walk:	Strides: 5–7 in. (12.7–17.8 cm)	
Trot:	Strides: 8–13 in. (20.3–33 cm)	
	Trail widths: 1¼–2⅜ in. (3.2–6 cm)	
Side trot:	Strides: 10–14 in. (25.4–35.6 cm)	
	Group lengths: 3–4¾ in. (7.6–12.1 cm)	
Straddle trot:	Strides: 14–16 in. (35.6–40.6 cm)	
Lope:	Strides: 5–9 in. (12.7–22.9 cm)	
	Group lengths: 15–19¾ in. (38.1–50.2 cm)	
Gallop:	Strides: 8–36 in. (20.3–91.4 cm)	
	Group lengths: 25–45 in. (63.5–114.3 cm)	

NOTES:

- Explore in a rocking-horse lope, as well as use side- and direct-register trots when traveling. They transition into walks when they slow down to make or investigate scent marks, or to forage.

- In some parts of their range, such as Bakersfield, California, Kit Foxes are doing very well in urban environments.

- Use underground dens and burrows throughout the year to rest, escape predators, and rear kits.

SCATS: $^{3}/_{16}$–⅝ in. (0.5–1.6 cm) diameter, 2–4½ in. (7.6–11.4 cm) long

Scats of hair and bone are twisted and tapered ropes with pointy ends. Kit Foxes are almost completely carnivorous. Kit Foxes post scats on elevated surfaces along trails or roads, and form latrines near the base of prominent shrubs or other visual features in their range.

URINE AND OTHER SCENT-MARKING BEHAVIORS: Similar to the urine of Red Foxes, that of Kit Foxes has a powerful skunklike odor. At times this smell is nearly on a par with that of Red Foxes. They urinate on any elevated surface that might hold and convey scent.

In Swift Foxes, a close relative of the Kit Fox, single scats and latrines are found in higher densities in the *core areas* of a pair's home range (that part of their territory that does not overlap with other foxes). In addition, latrines are also found in higher densities in areas where a pair's home range overlapped with neighboring foxes, suggesting that latrines are used in territorial defense as well as like "bulletin boards" for information exchange (Darden et al. 2008). Darden, Steffensen, and Dabelsteen also report that in about 25 percent of latrine visits, foxes only sniff the area, and when they do themselves scent-mark, it is more often with urine than with fecal matter.

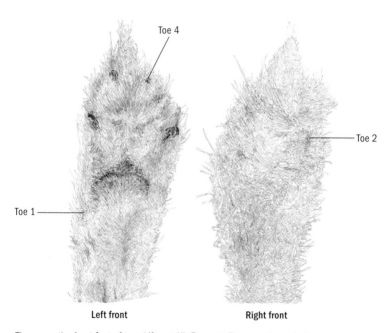

Toe 4

Toe 2

Toe 1

Left front   Right front

These are the front feet of two different Kit Foxes to illustrate the variation in the amount of fur covering the pads.

Dogs: Family Canidae

Trotting Kit Fox.

Side trot of a
Kit Fox.

Galloping
Kit Fox.

Dainty Kit Fox tracks in rare mud in Death Valley National Park. Some Kit Foxes exhibit soles nearly completely covered in fur, as seen here.

Kit Fox tracks in sand; the larger is the front track.

RED FOX

Vulpes vulpes

Plate 23
Plate 109

TRACKS: Front: 1⁷/₈–2⁷/₈ in. long (4.8–7.3 cm) × 1³/₈–2¹/₈ in. wide (3.5–5.4 cm); young 1³/₈ × 1³/₈ in. (3.5 × 3.5 cm)

Small to medium. Digitigrade. Symmetrical. Five toes. Toe 1 is greatly reduced and raised up on the inner leg. Four toes reliably register in tracks. The metacarpal pads are fused and create one triangular pad. The marks from the digital and metacarpal pads are greatly reduced by fur growing on the feet, though less so during summer months; the negative space between the toes and pads appears exaggerated by the excess fur and reduced exposure of toe pads. The metacarpal pad often appears in the track as a chevron-shaped or flat line, and is a key feature for quick identification. The negative space between toes and metacarpals forms an "X." There is also an additional carpal pad higher on the leg that registers when the fox is pouncing or at high speeds. Claws are semiprotractible and may or may not register. Front tracks are larger and often rounder than hind tracks.

Hind: 1⅝–2½ in. long (4.1–6.4 cm) × 1¼–1⅞ in. wide (3.2–4.8 cm); young 1⅜ × 1³⁄₁₆ in. (3.5 × 3 cm)

Small to medium. Digitigrade. Symmetrical. Four toes. The metatarsal pads are fused and create one triangular pad; the metatarsal pads may only partially register. The marks from the digital and metacarpal pads are greatly reduced by fur growing on the feet, especially during winter months; the negative space between the toes and pads appears exaggerated by the excess fur and reduced exposure of toe pads. The negative space between toes and metacarpals forms an "X." Claws are semiprotractible and may or may not register.

TRAIL:

Walk:	Strides: 8–12 in. (20.3–30.5 cm)
Trot:	Strides: 13–20 in. (33–50.8 cm) Trail widths: 2–3¾ in. (5.1–9.5 cm)
Side trot:	Strides: 14–22¾ in. (35.6–57.8 cm) Group lengths: 4½–8 in. (11.4–20.3 cm)
Lope:	Strides: 9–15 in. (22.9–38.1 cm) Group lengths: 20–29 in. (50.8–73.7 cm)
Gallop:	Strides: 8–55 in. (20.3–139.7 cm) occasionally up to 109 in. (276.9 cm) Group lengths: 25–60 in. (63.5–152.4 cm)

NOTES:

• There are two converging groups of Red Foxes in California. The first is native and occurs in the Sierra Nevadas and the northern Sacramento Valley. The second is an exotic population released in several areas along the coast, and subsequently spreading east into the Sacramento Valley. The distinction between these two populations is currently under study.

• Nearly always travel in a direct-register trot, although they also frequently use a side trot to cover ground more quickly. They will occasionally use a straddle trot, but never for sustained distances. Red Foxes also walk to investigate or forage, and use lopes and gallops at speed.

• In deep snow they most often walk, but they will also bound, where each hole in the snow is all four feet and body.

• Although Red Foxes do not climb, they easily move up angled trunks, and will travel along the tops of stone walls, logs, and other narrow, elevated surfaces; they have a catlike agility.

SCAT: ⁵⁄₁₆–¾ in. (0.8–1.9 cm) diameter, 3–6 in. (7.6–15.2 cm) long

Neither color nor size alone provides the necessary information to distinguish Gray from Red fox scats reliably; both species eat fruits, mast, and meat. Scats of fruit and mast crops are tubular and typically of a larger diameter and volume than meat scats; fruit scats exhibit little to no twisting, a smooth surface, and have either blunt or tapered ends. Scats of hair and bone are twisted and tapered ropes with pointy ends.

Foxes post scats in obvious places, often elevated, and in the middle of trails or roads. Trail junctions are heavily marked as well, where they may accumulate over time. Scats may also accumulate along regular travel routes, and near regular bedding sites and dens.

URINE AND OTHER SCENT-MARKING BEHAVIORS: The smell of Red Fox urine is incredibly pungent and powerful, and is often described as being similar to Striped Skunk spray. During the mating season, clouds of scent wafting through forests can be successfully backtracked to scent posts up to 50 yards away. Like other canids, Red Foxes urinate on any elevated surface that might hold and convey scent.

J. D. Henry (1993), a renowned fox researcher, reports that Red Foxes use any of 12 postures to urinate, depending upon where they are trying to leave a scent mark. Males don't always lift a leg, and females don't always squat. Males do, however, tend to urinate much farther in front of the rear feet than females, while females, when squatting, drop urine almost right between the rear tracks.

Red Foxes also dribble urine as they scavenge (Henry 1993), and mark exhausted food caches (Henry 1977), which researchers believe may aid them in identifying areas they have already searched and therefore increase their foraging efficiency.

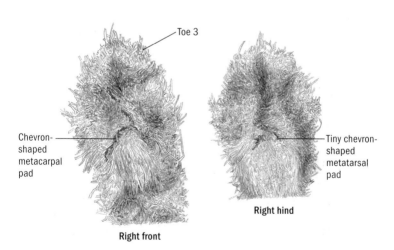

Toe 3

Chevron-shaped metacarpal pad

Tiny chevron-shaped metatarsal pad

Right hind

Right front

Red Fox feet.

Direct-register trot
of a Red Fox.

Side trot of a
Red Fox.

Perfect front track of a Red Fox along the Sacramento River reveals the details of the fur and the pebbly toe and metacarpal pads peeking through.

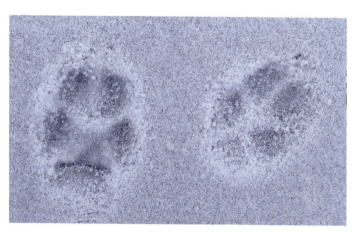

Red Fox tracks in snow. Look carefully for the chevron-shaped impression made by the palm pad in the front track on the left.

The scat of a Red Fox deftly placed atop a pine seedling.

A Red Fox pauses long enough to dribble urine on a projecting tip of wood. The snow makes it much easier to see the stance of the animal and the urine.

Bears: Family Ursidae

AMERICAN BLACK BEAR
Ursus americanus

Plate 34

Plates 117, 124, and 125

TRACKS: Front: 3¾–8 in. long (9.5–20.3 cm) × 3¼–6 in. wide (8.3–15.2 cm); cub 3¼ × 2¾ in. (8.3 × 7 cm)

Large to very large. Plantigrade. Asymmetrical. Five toes. Toe 1 is the smallest and is located on the inside of the track, and does not register reliably. The metacarpal pads are fused and create one large curved pad, much wider than long. There is an additional circular carpal pad (the heel), at the posterior edge of the track that may or may not register. The negative space between the toes and metacarpals is filled with fur, which at times blurs the overall track appearance. Claws may or may not register, and vary in length from bear to bear; the claws on the front feet are longer than those on the hind feet.

Hind: 5³⁄₈–8⁷⁄₈ in. long (13.7–22.5 cm) × 3½–6 in. wide (8.9–15.2 cm); cub 3¼ × 2¾ in. (8.3 × 7 cm)

Large to very large. Plantigrade. Asymmetrical. Five toes. Toe 1 is the smallest. The metatarsal pads are fused and create one large pad. The "palm" and "heel" are covered in tough skin and are connected in the footprint, giving the track an appearance similar to a human footprint. There is a sideways "V" filled with fur at the place where the palm and heel meet in the track. The negative space between the toes and metatarsals is filled with fur, which at times blurs the overall track appearance. Claws may or may not register.

TRAIL:

Direct-register walk:
Strides: 17–25 in. (43.2–63.5 cm)
Trail widths: 8–14 in. (20.3–35.6 cm)

Overstep walk:
Strides: 19–28 in. (48.3–71.1 cm)
Trail widths: 8–14 in. (20.3–35.6 cm)

Trot:
Strides: 27–37 in. (68.6–94 cm)
Trail widths: 6–10 in. (15.2–25.4 cm)

3× Lope:
Strides: 25–30 in. (63.5–76.2 cm)
Group lengths: 38–50 in. (96.5–127 cm)

Gallop:
Strides: 24–60 in. (61–152.4 cm)
Group lengths: 49–75 in. (124.5–190.5 cm)

NOTES:

• Use an overstep walk when traveling. Walk about the landscape but trot to cover open ground, or when they feel threatened or exposed. They lope when startled, and gallop in extreme circumstances or when very afraid.

• Play in isolation as well as with other bears.

• Competent climbers, and will walk along logs and porch railings, balancing on narrow surfaces at various heights off the ground.

- Direct-register walk in deep snow, leaving a characteristic pigeon-toed trail, and will sometimes use a 2 × 2 lope, much like weasels.
- In California not all populations of black bears hibernate during the winter, and those in the far south and coastal ranges may only hibernate very short periods when weather is mild and food abundant.

SCAT: 1¼–2½ in. (3.2–6.4 cm) diameter, 5–12 in. (12.7–30.5 cm) long, sometimes larger. A cub's scat can be as small as ½ in. (1.3 cm) in diameter and easily confused with foxes or raccoons.

Black bears defecate between six and eight times each day (Fair and Rogers 1990). Bear scat varies tremendously, depending upon diet. Acorn scats often resemble light-brown patties or swirls of soft-serve ice cream that quickly turn black in open air. Fruit scats can be amorphous and loose, or tubular and filled with seeds. Fruit scats break easily and have blunt ends. Meat and internal-organ scats are loose patties, or very amorphous, and do not last long on the landscape. Scats composed of fur and bone, or fibrous roots (such as lily roots) are often linked segments.

Scats accumulate at regular bedding sites, near large kills or carcasses, or other food-abundant areas. However bear scats are also found sprinkled along their travel routes and appear to play a less prominent role than urine in chemical communication.

URINE AND OTHER SCENT-MARKING BEHAVIORS: Bears use urine to scent-mark home ranges and communicate. Females dribble urine nearly constantly as they go, and these chemical cues are thought to strengthen their territorial boundaries and exclude other adult females from their range. Bears also straddle and walk over saplings to deposit urine along travel routes; the females do this by first dribbling urine down the long hairs attached to their vulvas and then dragging them over vegetation (Fair and Rogers 1990).

Terry DeBruyn (1999) noted that the most common scenting behavior in the bears in his study group was straddling trees and urinating on them.

Black bears rub, bite, and claw trees to communicate with other bears (Grinnell, Dixon, and Linsdale 1937; Green and Mattson 2003). On larger trees they may rub the base with their flanks in passing, or pause long enough to bite the exposed roots as they rub. More often they stand to rub, claw, and bite the main trunk. Rubbing is exhibited whether they claw or bite the tree, and you can find their hairs stuck in the bark, sap, or along the lowest branches of trees they mark if you look closely. Burst and Pelton (1983) noted that black bears in the Smoky Mountains marked trees with a much higher frequency during the months of May through July than in the months of August through October. Marking peaked in midsummer during breeding season. They also found that 74 percent of marked trees were found within 15 m (about 17 yards) of a trail, and 75 percent of marked trees were on the downhill side of the trail. Bears also mark telephone poles, signs, front porches, and other human objects in just the same way: rubbing, biting, and clawing.

Black bears also decapitate evergreen saplings and other trees, sometimes climbing 40 ft (12 m) up to remove the top 3 ft (1 m) of a large tree (Elbroch 2003). They bite through them with their teeth while pulling the top with their forelimbs.

There has been conjecture that the height at which a bear marks conveys something about dominance to other bears, but there is actually no evidence to support this hypothesis. In fact, bears are frequently seen rolling on their favorite rubbing posts when they tip over. Black bears also sometimes climb to mark higher on the tree, especially younger, smaller bears—and young bears hold the lowest social rank, not the highest.

Bears that encounter a tree marked by another bear are most likely to over-mark it if it was made by a bear of the same sex. Males overmark male scents, and females overmark female scents (Kilham and Gray 2003).

Bears also scent-mark with their feet. Along certain sections of trails, bears step in exactly the same spots as their predecessors, and over time zigzags of deep, worn circles appear in the debris or ground. With stiff, straight front legs, they deliberately emphasize the normal twist of their front feet while they walk, as if rubbing scent into the earth. Some bears walk through such a section of trail and then loop back to do it again (Kilham and Gray 2003). Short sections of "bear trails" in which they follow in each other's footsteps are often associated with wallows, mark trees, and abundant food resources. Occasionally long trails of these steps emerge, traveling over ridges and between bedding and feeding areas.

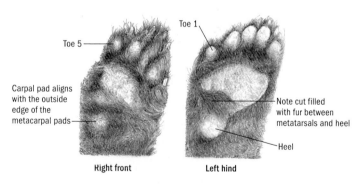

American Black Bear feet.

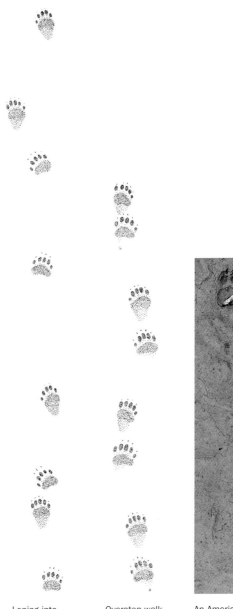

Loping into galloping American Black Bear.

Overstep walk of a American Black Bear.

An American Black Bear in an overstep walk—front, hind, front, hind.

The copious fur on this American Black Bear's feet registered very well in this silty wash.

The characteristic signs of a tree bitten by a marking American Black Bear.

The top of this evergreen tree has been bitten and ripped off by a marking bear.

The worn-in circles in the vegetation denote a well-used and marked bear trail.

Sealions: Family Otariidae

CALIFORNIA SEALION
Zalophus californianus

Plates 116 and 123

TRACKS: Front: 8–14 in. long (20.3–35.6 cm)

Very large. Asymmetrical. Five toes, although not all toes reliably register in tracks. Palm is unique to aquatic lifestyle, and in contrast to harbor seals that drag themselves on land, register very clearly; on hard substrates the deep impressions made by the palms may be all that is easily interpreted. Body drag is a significant characteristic of trails. Claws are sharp and apparent, and when present, are useful to distinguish their trails from those of larger sea turtles.

NOTES:

- Sealions can walk in an upright position on their front limbs and hind feet. This movement is in stark contrast to the seals, which drag their entire bodies along the ground as they "inchworm" their way to and from loafing areas.

- Sealions are gregarious, and where you find one you'll often find others. They also can be found with other pinniped species in large colonies or roosts.

SCAT: Tubular scats: 1–2 in. (2.5–5.1 cm) diameter

Pinniped scats are typically amorphous and pattylike when they are feeding upon fish and squid. They are tubular when they eat crabs and shellfish, and occasionally tubular when eating certain types of fish. Scats can be common at haulouts, but soft scats are swept up by the sea and/or weather and deteriorate quickly.

A sealion trail in firm sand. Rather than the paired set of claws characteristic of Harbor Seals, look for the impressions of the rounded "heel" of the front flippers to register in zigzagging patterns.

Seals: Family Phocidae

HARBOR SEAL *Phoca vitulina*

Plate 35
Plate 123

TRACKS: Front: 8–12 in. long (20.3–30.5 cm) × 7–10 in. wide (17.8–25.4 cm)

Very large. Asymmetrical. Five toes, although not all toes reliably register in tracks. Palm is unique to aquatic lifestyle. Body drag is a significant characteristic of trails. Claws are sharp and apparent, and are useful to distinguish their trails from those of larger sea turtles.

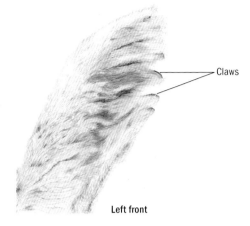

Harbor Seal front flipper.

Claws

Left front

The typical trail of a Harbor Seal on firm wet sand, with paired tracks and prominent claw marks. Compare these with the trail of the sealion in firm substrate, where the wrist is the predominant feature—since sealions move in an upright position.

The beautiful trail of a Harbor Seal, with characteristic claw marks that help differentiate them from sea turtles.

NORTHERN ELEPHANT SEAL *Mirounga angustirostris*

TRAIL: Trail width: Up to 6 ft (182.9 cm) across

NOTES:

• Seals are marine mammals, yet can be seen following estuaries or even upstream of freshwater outlets.

• Seals are gregarious and you'll often find them in groups. They also can be found with other pinniped species in large colonies or roosts.

SCAT: Scats of pinnipeds are most often amorphous and pattylike when they are feeding upon fish and squid. Scats are tubelike when they eat crabs and shellfish, and occasionally also when eating certain types of fish. Scats can be common at haulouts, but soft scats are swept out to sea, and/or they deteriorate quickly.

Weasels: Family Mustelidae

SEA OTTER *Enhydra lutris*

Plates 116 and 119

TRACKS: Front: 1¾–3¹/₈ in. long (4.4–7.9 cm) × 2–2³/₈ in. wide (5.1–6 cm). Data compliments of Max Allen.

Medium. Plantigrade. Round to oval. Five toes. The toes are fused together and are sometimes difficult to differentiate from each other in tracks; toes tend to form one larger C-shaped depression separate from the metacarpal pads. The metacarpal pads are fused and form one large oval depression bigger than the depression made by the fused toes; the carpal pad (the heel), at the posterior edge of the track, may or may not register. Claws may or may not register. The front track is very round, and the track never splays. The front feet are tiny when compared with the hind tracks; the huge discrepancy in size between the front and hind tracks is a key to Sea Otter track identification.

Hind: 6–8¹/₈ in. long (15.2–20.6 cm) × 3⁷/₈–4¾ in. wide (9.8–12.1 cm). Data compliments of Max Allen.

Large to very large. Plantigrade. Highly asymmetrical. Five toes. The toes are somewhat fused into flippers like those of seals and sealions, but digital pads are typically evident. The metatarsals are fused to form one very large pad. Claws often register. The heel of the hind foot often registers, and the track is reminiscent of a human footprint; however, the concave "arch" is on the outside of their feet rather than the inside. Hind tracks are huge in comparison to the front tracks.

NOTES:

• Sea Otters are marine mammals, and their tracks and signs are relatively rare along California's beaches.

- Walk on land, and their tiny front feet register much closer to the midline of the trail than their massive hind tracks, which straddle their front ones; the resulting trail is highly distinctive.
- Bound and lope in patterns reminiscent of Spotted Skunks, meaning that they are boxy and highly variable.

SCAT: Tubular scats: $^7/_8$–1½ in. (2.2–3.8 cm) diameter. Data compiled from Murie and Elbroch (2005) and Max Allen (unpublished data).

Sea Otter scat is very similar to river otter scat, but on a larger scale. They create tubular scats when feeding on urchins and crayfish, and dark, amorphous patties filled with fish scales when eating fish. Like river otters, Sea Otters also defecate black, tarry liquid scats, and viscous, yellow liquid scats. We do not know whether these are waste products from eating animals like squid or a cleansing of the intestinal tract, as is suspected for river otters.

Sea Otters often defecate in the water, but occasionally they use regular sites on rocks or stretches of beach where they come ashore to defecate, forming large latrines over time. Scats on beaches are often found in association with digs and rolls, a behavior also observed in river otters utilizing marine environments.

URINE AND OTHER SCENT-MARKING BEHAVIORS: Sea Otters lack anal scent glands (Tarasoff 1972); the aquatic lifestyle of this mammal prioritizes vocal over chemical communication.

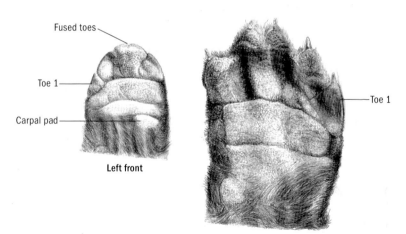

Sea Otter feet.

Walking Sea Otter.

A set of four Sea Otter tracks, from left to right, the left hind, left front, right front, and right hind. Photo by Max Allen.

The peculiar walking trail of a Sea Otter. The small round holes following the midline of the trail are the front feet. Photo by Max Allen.

A dig, roll, and soft scat of a Sea Otter along California's coast near Santa Cruz. Photo by Max Allen.

NORTH AMERICAN RIVER OTTER *Lontra canadensis*

Plate 29

Plates 111 and 119

TRACKS: Front: 2¹⁄₈–3¼ in. long (5.4–8.3 cm) × 1⁷⁄₈–3 in. wide (4.8–7.6 cm)

Medium. Plantigrade. Asymmetrical. Five toes. Toe 1 is the smallest toe. Webbing may or may not show. The metacarpal pads are fused and strongly lobed; the carpal pad (the heel), at the posterior edge of the track, may or may not register. The negative space between toes and metacarpals has little to no fur. Claws may or may not register. The front track is more symmetrical than the hind track, and smaller in dimensions; this size comparison between front and hind feet is useful in differentiating the trails of similar-sized river otters and Fishers.

Hind: 2⅛–4 in. long (5.4–10.2 cm) × 2⅛–3¾ in. wide (5.4–9.5 cm)

Medium. Plantigrade. Asymmetrical. Five toes. Toe 1 is well developed and very long. Mesial webbing registers reliably. The metatarsals are fused to form one large pad. Claws may or may not register. Hind tracks are larger than front tracks. Because toe 1 is situated farther back in hind tracks than front tracks, the hind tracks appear more asymmetrical than the front tracks. When toe 1 is not visible, the hind tracks appear catlike and symmetrical.

TRAIL:	Walk:	Strides: 5¾–14 in. (14.6–35.6 cm)
		Trail widths: 4½–7 in. (11.4–17.8 cm)
	3 × 4 lope:	Strides: 6–28 in. (15.2–71.1 cm)
		Group lengths: 10–20½ in. (25.4–52.1 cm)
	2 × 2 lope:	Strides: 15–40 in. (25.4–101.6 cm)
		Trail widths: 4½–7 in. (11.4–17.8 cm)
		Group lengths: 6–12½ in. (15.2–31.8 cm)
	Slide:	Trail widths: 6–10 in. (15.2–25.4 cm)

NOTES:

- Some tracks can easily be confused with those of raccoons. In raccoons the toes are flat and tend to connect all the way to the metacarpal/metatarsal pad, while in otters the toes are curved so that the toe pads and the metacarpal/metatarsal pads register more so than the middle portion of the toes. Tracks may also be confused with Fisher tracks, and this comparison is discussed under the Fisher account.

- Travel in a 3 × 4 lope, and explore in a walking gait.

- Inhabit all sorts of wetland and water habitats, including saltwater marshes, open beaches, and tidal pools. However, otters also cross large land masses, and even mountains, while moving between water systems.

- Slide in varied substrate, from leaf litter to snow. Downhills or flat ground will always be traversed with gallops and slides in snow, and they will even slide short distances uphill.

- Use culverts and drains, as well as investigate and rest in burrows and cavities in root systems of all kinds.

- Form runs between water systems, especially evident at the narrowest point between two water sources.

- Unlike other mustelids, otters are often found in pairs or in family and bachelor groups.

SCAT: Tubular scats: ⅜–1 in. (1–2.5 cm) diameter, 3–6 in. (7.6–15.2 cm) long

Scats vary from amorphous, pattylike squirts to tubular structures but always reek of fish and disintegrate into a pile of fish scales. Scats of crayfish are more tubular and pink or red. Otters often feed primarily on crayfish in times of abundance. A study in Georgia found crayfish in 98 percent of summer scats (Noordhuis 2002).

Scats are often deposited in larger latrines near large prominent trees near water sources, near holes in banks where they enter and exit the water, at crossing points where otters travel across land between two bodies of water, or on peninsulas jutting out onto water sources. Like all mustelids, otters mark elevated surfaces, such as stumps, as well as bridges, logs, or other surfaces over water or forest debris. In coastal areas where otter densities are high, latrine sites up to 65 ft (20 m) across dot the coast every 1,000 ft (300 m) (Ben-David et al. 1998). More often, they are about 11 ft^2 (1 m^2) and placed in prominent locations along an inhabited body of water. Traveling otters mark rolling sites and latrines as they are encountered. Two animals in Idaho marked eight latrine sites with scats along 1.5 miles (2.4 km) of stream bank (Melquist and Hornocker 1983).

In Alaskan coastal populations there are both social and solitary river otters foraging in very different ways. Social otters are members of larger bands and defecate intensively in a smaller number of latrines to remain in contact with pack members, so that they can reunite when they want to. Nonsocial animals travel singly or in small familial groups and defecate in a much higher number of sites, but less frequently within each. Nonsocial otters probably employ scent marks to maintain mutual avoidance territoriality (Ben-David et al. 2005).

URINE AND OTHER SCENT-MARKING BEHAVIORS: River otters also produce two additional glandular secretions at latrine sites that might be confused with scats. They slough a yellow-greenish jelly from the intestines and a white, gooey substance, which some believe is from the intestinal tract, and others believe is from the anal glands.

Some researchers call otter latrines "rolls," because rolling and digging also often occur at these sites. Some researchers believe rolling is necessary to maintain oils in the coat, thus keeping the animals warm and waterproof. Debris or earth may be mounded, and scats are often placed on top or nearby. Otters also create scrapes in mud and sand, which look somewhat catlike, and scats may appear in or near them. Where otters are often active, entire forest floors may be swept into neat piles scattered at intervals near the shoreline. In addition, otters collect vegetation and create circular globs or mounds, upon which to scent-mark, as well as twist and form grasses and vegetation that is still rooted, which is also scented with urine and anal glands. When an otter urinates, the urine stream jets forward if it is a male and backward if it is a female.

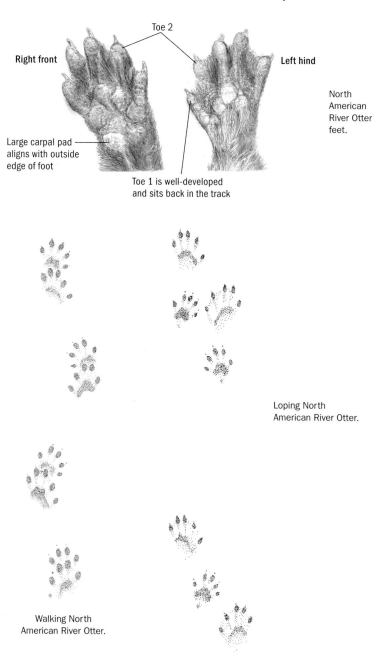

Toe 2

Right front

Left hind

North American River Otter feet.

Large carpal pad aligns with outside edge of foot

Toe 1 is well-developed and sits back in the track

Loping North American River Otter.

Walking North American River Otter.

On the right side of the photo, the understep walk of a river otter along the Sacramento River. In the upper left corner of the photograph are the tracks of a raccoon, and if you look closely a Ringtail is also crossing from left to right.

The loping and sliding trails of river otters across a frozen lake surface.

A river otter roll. Note the impression of the body as well as accompanying urine and scat.

Chewed grass at a river otter marking site along a river bank.

WOLVERINE   *Gulo gulo*

Plate 31
Plate 112

TRACKS: Front: 3⅝–6¼ in. long (9.2–15.9 cm) × 3½–5¼ in. wide
(8.9–13.3 cm)

Medium to very large. Plantigrade. Asymmetrical. Five toes. Toe 1 is
the smallest, located on the inside of the track. The metacarpal pads
are fused, yet strongly lobed; the carpal pad (the heel), at the posterior edge of the track may or may not show. The negative space in
between the toes and the metacarpals are especially furry in winter,
and blur the appearance of the track. Claws may or may not register.
The front track is more symmetrical than the hind track, but front
and hind track parameters are similar.

Hind: 3⅝–6 in. long (9.2–15.2 cm) × 3¼–5¼ in. wide (8.3–13.3 cm)

Medium to large. Plantigrade. Asymmetrical. Five toes. Toe 1 is
the smallest and does not register reliably. The metatarsal pads are
fused, yet strongly lobed. The furred "heel" often registers as well.
The negative space between toes and metatarsals is furry, and may
influence the track appearance. Claws may or may not register.
When toe 1 is not visible, the hind tracks appear very symmetrical
and at a glance can easily be confused with tracks of Domestic Dogs,
Cougars, or Wolves.

TRAILS: 3 × 4 Lope: Strides: 7–45 in. (17.8–114.3 cm)
Group lengths: 27–45 in. (68.6–114.3 cm)

Walk: Strides: 8–15 in. (20.3–38.1 cm)

2 × 2 Lope: Strides: 12–40 in. (30.5–101.6 cm)
Trail widths: 7–10 in. (17.8–25.4 cm)

NOTES:

- There was a confirmed case of a Wolverine just north of Tahoe in 2008, the first such confirmation in decades; the animal was also documented in 2009 and 2010.

- Explore and hunt in a lope, traveling great distances. They slow down to a walk to investigate potential food sources and to engage in scenting and other behaviors.

- Comfortable in the open, often seen moving at higher altitudes above tree line, as well as inhabiting the treeless tundra in the far North.

- Competent climbers.

- In deep snow Wolverines either walk or use a 2 × 2 lope, and slide frequently in steep terrain.

- In all weasel species toe 1 tends to register the lightest. In cases where only four toes are visible in the track, tracks can be confused with those of canids or felids of similar size.

SCAT: $^3/_8$–1 in. (1–2.5 cm) diameter, 3–8 in. (7.6–20.3 cm) long

Fur and bone scats are twisted, tapered ropes with pointed ends. Longer scats often fold over upon themselves. Wolverines mark elevated surfaces, such as stumps, as well as bridges, logs, and boulders.

URINE AND OTHER SCENT-MARKING BEHAVIORS: Wolverines roll on other animals' trails and scent posts to overmark them. Like Fishers, they also roll on, climb, bite, and "beat up" small saplings, breaking limbs and defoliating them. They also climb and regularly mark larger trees along travel routes, even going out of their way to return to them when nearby, where they climb, rub, and claw the bark; sometimes they mark and claw the bark while hanging upside down (Koehler et al. 1980).

The clean, front right footprint of a Wolverine.

Loping
Wolverine.

The beautiful fresh trail of a large walking Wolverine. Photo by Alex Huryn.

AMERICAN MARTEN *Martes americana*

Plates 7 and 8

Plate 23

Plates 103 and 104

TRACKS: Front: 1⅝–2¾ in. long (4.1–7 cm) × 1½–2⅝ in. wide (3.8–6.7 cm)

Small to medium. Plantigrade. Asymmetrical. Five toes. Toe 1 is the smallest, located on the inside of the track, and does not register reliably. Several metacarpal pads are fused, and others register individually; there are two additional pads (the heel) at the posterior edge of the track that may or may not show. The negative space between the toes and metacarpals is filled with fur, which typically blurs the overall track appearance. Claws are fine, and may or may not register. The front track is more symmetrical than the hind track.

Hind: 1½–2¾ in. long (3.8–7 cm) × 1½–2¼ in. wide (3.8–5.7 cm)

Small to medium. Plantigrade. Asymmetrical. Five toes. Toe 1 is the smallest and often overlooked in tracks. Some of the metatarsal pads are fused, and others register separately. The furred "heel" often registers, as well. The negative space between the toes and metatarsals is filled with fur, which at times blurs the overall track appearance. Claws may or may not register. Because toe 1 is situated farther back in hind tracks than front tracks, the hind tracks appear more asymmetrical than the front tracks. When toe 1 is not visible, the hind tracks appear symmetrical and foxlike.

TRAIL:

3 × 4 Lope:	Strides: 8–32 in. (20.3–81.3 cm)
	Group lengths: 8½–20 in. (21.6–50.8 cm)
2 × 2 Lope:	Strides: 10–34 in. (25.4–86.4 cm),
	occasionally up to 72 in. (182.9 cm)
	Trail widths: 2½–4½ in. (6.4–11.4 cm)
	Group lengths: 3½–7½ in. (8.9–19.1 cm)
Walk:	Strides: 5–9 in. (12.7–22.9 cm)
	Trail widths: 3¼–5½ in. (8.3–13.3 cm)
Body print (without tail):	10–14 in. (25.4–35.6 cm)
	long × 4–8 in. (10.2–20.3 cm) wide

NOTES:

• Patrol their home ranges in a 3 × 4 lope (a 2 × 2 lope is used in deep-snow conditions), exploring holes and nooks within root wads and boulder jumbles. They slow to a walk to investigate or forage on the ground.

• Martens are generally more comfortable in open habitats than are Fishers.

• Use trails readily, and mark them more than Fishers do. They will cross roads but often seek culverts to do so.

• Competent climbers, and they invest a great deal of time hunting in the trees. Their hind feet can turn 180 degrees, allowing them to move head first down trunks. Look for body prints in deep snow, where they have jumped from a tree rather than climbed down; body prints are found more often along marten trails than those of Fishers.

• Toe 1 in all weasel species tends to register the lightest. In cases where only four toes are visible in the track, tracks can be confused with those of canids or felids of similar size.

SCAT: ³⁄₁₆–⅝ in. (0.5–1.6 cm) diameter, 2–5 in. (5.1–12.7 cm) long

Fruit and mast scats are tubular and of a larger diameter than meat scats; they typically exhibit smooth surfaces, lack twisting, and often have pointed ends. Fur and bone scats are extremely twisted, tapered ropes with pointed ends. Longer scats often fold over upon themselves. Martens mark elevated surfaces, such as stumps, as well as bridges, logs, or other surfaces over water or forest debris, and are especially prone to mark human hiking trails (which is unusual for Fishers). Scats also accumulate at the entrances of den sites, as well as in the latrines of their respective prey species.

URINE AND OTHER SCENT-MARKING BEHAVIORS: Martens urinate on elevated surfaces along travel routes, as well as drag a glandular area on their abdomen over logs and branches to scent mark (Clark et al. 1987). When they do this they spread their hind legs enough for the stomach and anus to touch the ground, and then use their front limbs to drag the body forward.

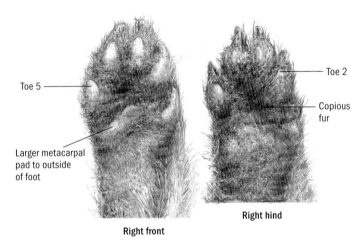

Toe 5

Larger metacarpal pad to outside of foot

Toe 2

Copious fur

Right hind

Right front

American Marten feet.

Loping American
Marten.

Walking American
Marten.

The 2 × 2 lope of an
American Marten.

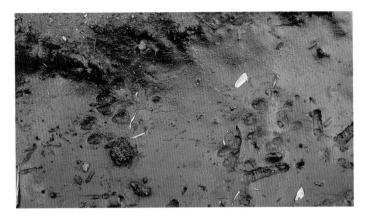

Perfect American Marten tracks in mud, along with accompanying grouse tracks in the lower right corner.

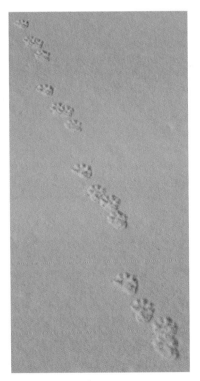

The beautiful loping gait of an American Marten atop firm snow near Tahoe.

FISHER

Martes pennanti

Plate 9
Plate 29
Plate 106

TRACKS: Front: 2¹⁄₈–3⁷⁄₈ in. long (5.4–9.8 cm) × 1⁷⁄₈–4¼ in. wide (4.8–10.8 cm)

Medium. Plantigrade. Asymmetrical. Five toes. Toe 1 is the smallest, located on the inside of the track, and does not register reliably. Several metacarpal pads are fused, and others register individually; there are two additional pads (the heel) at the posterior edge of the track, of which one may or may not show. The negative space between the toes and metacarpals is filled with fur, which at times blurs the overall track appearance. Claws are fine and may or may not register. The front track is more symmetrical than the hind track.

Hind: 2–3¹⁄₈ in. long (5.1–7.9 cm) × 1½–3½ in. wide (3.8–8.9 cm)

Medium. Plantigrade. Asymmetrical. Five toes. Toe 1 is the smallest, and may or may not register. Some of the metatarsal pads are fused, and others register separately. The furred "heel" often registers as well. Fur fills the negative space between the toes and metatarsals, and this sometimes blurs the overall track appearance. Claws may or may not register. Because toe 1 is situated farther back in the hind tracks than in the front tracks, the hind tracks appear more asymmetrical than the front tracks. When toe 1 is not visible, the hind tracks appear Bobcat-like and more symmetrical.

TRAIL:

3 × 4 Lope:	Strides: 5¾–30 in. (14.6–76.2 cm) Group lengths: 12–26 in. (30.5–66 cm)
Full gallop:	Strides: 15–31 in. (38.1–78.7 cm) Group lengths: 31–40 in. (78.7–101.6 cm)
2 × 2 Lope:	Strides: 17–45 in. (43.2–114.3 cm), have seen up to 78 in. (198.1 cm) Trail widths: 3–5½ in. (7.6–14 cm) Group lengths: 6–13 in. (15.2–33 cm)
Walk:	Strides: 7–11½ in. (17.8–29.2 cm) Trail widths: 5–6½ in. (12.7–16.5 cm)
Trot:	Strides: 13½–18 in. (34.3–45.7 cm) Trail widths: 3–4⁷⁄₈ in. (7.6–12.4 cm)
Body print (without tail):	10–18 in. (25.4–45.7 cm) long

NOTES:

- Use a 3 × 4 lope to travel, and slow to a walk to explore holes and nooks within root wads and boulder jumbles, and to forage on the ground.

- Avoid moving in open areas more than martens do.

- Competent climbers, and spend time hunting in the trees as well as on the ground. However, large males spend considerably less time in trees than do the much smaller and lighter females, so much so that climbing itself is a decent indicator of the gender of the animal that made the trail that you are following (Powell 1993). Fishers can turn their hind feet up to 180 degrees, allowing them to move head first down trunks. Body prints in deep snow can indicate where they have jumped from a tree to the ground instead of climbing down.

- In deep snow, Fishers tend toward a 2 × 2 lope or walk.
- In all weasel species toe 1 tends to register the lightest. In cases where the track only shows four toes, tracks can be confused with those of canids or felids of similar size.
- Fishers and otters leave similar tracks. The hind tracks of otters are larger than their front tracks and are more obviously webbed; the front feet of Fishers are larger than their rear feet. The fur on the soles of Fisher feet is quite dense, while it is sparse on the feet of otters. For this reason the digital pads tend to be much smaller in Fisher tracks in proportion to the overall size of the tracks. Also, study the size of toe 1 on each hind foot. This toe is well developed on the otter, and the digital pad is equal to the other toes; toe 1 is also long in otters and registers far from the palm. In Fishers, toe 1 on the hind feet is reduced, the smallest digital pad by far, and often is absent in their tracks altogether.

SCAT: ³/₁₆–¾ in. (0.5–1.9 cm) diameter, 2–6 in. (5.1–15.2 cm) long

Fruit and mast scats are tubular and of a larger diameter than meat scats; they typically exhibit smooth surfaces, lack twisting, and often have pointed ends. Fur and bone scats are extremely twisted, tapered ropes with pointed ends. Longer scats often fold over upon themselves. Fishers mark elevated surfaces, such as stumps, as well as bridges, logs, or other surfaces over water or forest debris; however, unlike martens, they are less likely to mark human hiking trails (Elbroch 2003). Scats also accumulate at the entrances of den sites, as well as in the latrines of raccoons.

Fisher scats, possibly more so than those of other mustelids, can vary tremendously in size. Scats can be tiny, of a size easily confused with the scat of a weasel, or they can be large, similar in proportions to the scat of a fox.

URINE AND OTHER SCENT-MARKING BEHAVIORS: Fishers mark with urine and by rolling. They roll on other animals' trails and scent posts, including those of foxes. They roll on and "beat up" small saplings, breaking limbs and defoliating them (Elbroch 2003); look closely for hairs stuck in the foliage and stumps upon which they roll and rub.

Fishers also drag their bellies over small stumps, rocks, and mounds of snow (Powell 1981). Occasionally this form of marking leaves troughs in snow that measure up to 8 ft (2.4 m) in length.

During the breeding season Fishers frequently deposit black, tarry, fecal-like excretions (Powell 1981). Look for such sign in the same circumstances you find scats.

Weasels: Family Mustelidae

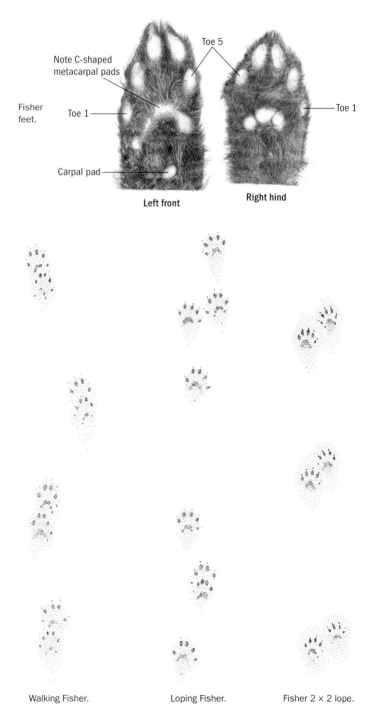

Note C-shaped metacarpal pads

Toe 5

Fisher feet.

Toe 1

Carpal pad

Left front

Toe 1

Right hind

Walking Fisher.

Loping Fisher.

Fisher 2 × 2 lope.

The larger front and smaller hind tracks of a large Fisher.

The trail created by a scent-marking Fisher looks like a trough in soft snow conditions.

Front and hind Fisher tracks in snow; the larger track with longer claws is the front track.

Hair stuck in a small dead sapling rubbed vigorously by a marking Fisher.

Weasels

Plate 2

Plates 13 and 15

Plates 101 and 102

ERMINE (SHORT-TAILED WEASEL, STOAT)   *Mustela erminea*

TRACKS: Front: $^7/_{16}$–$^5/_8$ in. long (1.1–1.6 cm) × $^7/_{16}$–$^5/_8$ in. wide (1.1–1.6 cm)

LONG-TAILED WEASEL (BRIDLED WEASEL)   *Mustela frenata*

TRACKS: Front: $^5/_8$–$1^7/_{16}$ in. long (1.6–3.7 cm) × ¾–$1^3/_{16}$ in. wide (1.9–3 cm)

Tiny to small. Plantigrade. Asymmetrical. Five toes. Toe 1 is the smallest, located on the inside of the track, and does not register reliably. Several metacarpal pads are fused but individually lobed, and others register individually; the carpal pad (the heel), at the posterior edge of the track, may or may not register clearly. The fur-filled negative space between the toes and metacarpals may blur the appearance of the track. Claws may or may not register. The front track is often larger and more symmetrical than the hind track; however, occasionally the hind tracks register larger than the front ones.

Ermine

Hind: $7/16$–$9/16$ in. long (1.1–1.4 cm) × $3/8$–$3/4$ in. wide (1–1.9 cm)

Long-tailed Weasel

Hind: $3/4$–$1\frac{1}{2}$ in. long (1.9–3.8 cm) × $9/16$–1 in. wide (1.4–2.5 cm)

Tiny to small. Plantigrade. Asymmetrical. Five toes. Toe 1 is the smallest. Some of the metatarsal pads are fused, and others register separately. The furred "heel" may or may not register. The negative space between toes and metatarsals is filled with fur that may blur the appearance of the track. Claws may or may not register. Because toe 1 is situated behind toes 2–5, the hind tracks appear more asymmetrical than the front tracks. When toe 1 is not visible, the hind tracks appear symmetrical.

TRAIL: 2 × 2 Gallop: Strides: 4–45 in. (10.2–114.3 cm)
Trail widths: $7/8$–$27/8$ in. (2.2–7.3 cm)

Long-tailed Weasel Only

Bound: Strides: 15–25 in. (38.1–63.5 cm)
Trail widths: $17/8$–$27/8$ in. (4.8–7.3 cm)
Group lengths: 6–11 in. (15.2–27.9 cm)

NOTES:

- In California there is some habitat segregation between the Ermine and Long-tailed Weasel. The Ermine is restricted to the Sierra Nevada, Cascade, and Coastal Ranges in northern California, while the Long-tailed Weasel is widespread throughout the state. Where the two species overlap, their tracks and trails are difficult to differentiate. Large male Ermine could easily be mistaken for small female Long-tailed Weasels. Use track width to analyze finer distinctions instead of trail width parameters, but at times you may not be able to differentiate between them with certainty.

- Weasels explore and hunt in a lope, or modified bound, but slow to a walk to explore potential resources. When very exposed, or in pursuit, weasels sometimes use a rabbitlike bound.

- The bounding and loping trails of weasels can be quite erratic, in that they are a mix of long and short strides; weasels can make sharp turns, and from one set of four tracks to the next, angle away in an almost complete circle.

- Ermine can climb; Long-tailed Weasels both climb and swim. It is not uncommon to find Long-tailed Weasel trails entering and exiting water.

- The trails of small weasels are forever exploring holes, nooks, and crannies in rock walls, root systems, under buildings, in trees, and under snow.

- In deep snow, small weasels tend to stick to a 2 × 2 lope but occasionally investigate in a walk. Weasels also tunnel and travel within deep, soft snow.

- On harder substrates, sometimes only the claws are visible in their tracks.

SCAT:

Ermine

$1/8$–$5/16$ in. (0.3–0.8 cm) diameter, $3/4$–$23/8$ in. (1.9–6 cm) long

Long-tailed Weasel

³/₁₆–³/₈ in. (0.5–1 cm) diameter, 1–3¼ in. (2.5–8.3 cm) long

Weasel scats vary depending upon diet. Fruit and mast scats are tubular and of a larger diameter than meat scats; they typically exhibit smooth surfaces, lack twisting, and often have pointed ends. Fur and bone scats are extremely twisted, tapered ropes with pointed ends. Longer scats often fold over upon themselves.

Weasels mark elevated surfaces, such as stumps, as well as bridges, logs, or other surfaces over water or forest debris. Weasels form impressive latrines near burrow entrances and in chambers in their burrows (Polderboer, Kuhn, and Hendrickson 1941). Latrines also often contain uneaten bones and skin (Pearce 1937).

URINE AND OTHER SCENT-MARKING BEHAVIORS: Pearce (1937) also observed a captive male weasel that would, after defecating, drag the anal region across the ground for 6–12 in. (15.2–30.4 cm).

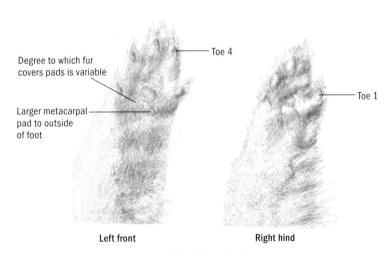

Left front   Right hind

Long-tailed Weasel feet.

The 2 × 2 lope
of an Ermine.

Mixed bounds and
lopes of a weasel.

Bounding Long-tailed
Weasel.

Typical for smaller weasels, this Ermine exhibits the partial tracks we are lucky to find in the best of substrates.

The heavy claws and rabbitlike bound of a Long-tailed Weasel.

Long-tailed Weasel tracks in which the hind prints have come in behind and overlapped atop the front tracks.

The accumulated scats in a weasel latrine immediately adjacent to its burrow entrance—the burrow in which it took up residence as a den.

AMERICAN MINK   *Neovison vison*

Plate 3
Plate 17
Plate 102

TRACKS: Front: 1¹⁄₈–1⁷⁄₈ in. long (2.9–7.3 cm) × ⁷⁄₈–1¾ in. wide (2.2–4.4 cm)

Small. Plantigrade. Asymmetrical. Five toes. Toe 1 is the smallest toe and located on the inside of the track. Proximal webbing may or may not show in the track. Several metacarpal pads are fused but lobed, and others register individually; there is an additional pad (the heel), at the posterior edge of the track, which may or may not show. The negative space between the toes and metacarpals contains less fur than the tracks of weasels, and mink also have larger, more bulbous metacarpal pads. Claws may or may not register. Front track is larger and more symmetrical than the hind track.

Hind: ¹³⁄₁₆–1¾ in. long (2.1–4.4 cm) × ¹⁵⁄₁₆–1⁵⁄₈ in. wide (2.4–4.1 cm)

Tiny to small. Plantigrade. Asymmetrical. Five toes. Toe 1 is the smallest toe and does not register reliably. Several metatarsal pads are fused but lobed, and others register individually. Proximal webbing may or may not show in the footprint. The negative space between the toes and metacarpals contains less fur than the tracks of weasels, and mink also have larger, more bulbous metacarpal pads. Claws may or may not register. Because toe 1 is situated behind toes 2–5, the hind tracks appear more asymmetrical than the front tracks. When toe 1 is not visible, the hind tracks appear symmetrical.

TRAIL: 2 × 2 Lope: Strides: 9–38 in. (22.9–96.5 cm) up to 50 in. (127 cm)
                              Trail widths: 2 –3¾ in. (5.1–9.5 cm)

            3 × 4 Lope: Strides: 6–36 in. (15.2–91.4 cm)
                              Group lengths: 5–14 in. (12.7–35.6 cm)

            Bound: Strides: 11–26 in. (27.9–66 cm)
                              Group lengths: 5–16 in. (12.7–40.6 cm)

            Walk: Strides: 3½–7¾ in. (8.9–19.7 cm)

            Slide: Trail widths 3–5 in. (7.6–12.7 cm)

NOTES:

• Explore and hunt in a 3 × 4 lope, or modified bound, but slow to a walk to explore potential resources. When exposed, or in pursuit, mink use a rabbit-like bound.

• In deep snow, mink tend toward the 2 × 2 lope. When using a 2 × 2 lope, mink make proportionately shorter and more consistent strides than weasels.

• Slide downhill in snow, but rarely on flat ground.

• Expert swimmers, and are typically associated with wetland systems and water bodies. However, they may be found traveling across large land expanses when hunting or moving between water sources.

• In some substrates, only teardrop-shaped toes and claws are visible in their tracks.

SCAT: ¼–⁷/₁₆ in. (0.6–1.1 cm) diameter, 1–4 in. (2.5–10.2 cm) long

            Fur and bone scats are extremely twisted, tapered ropes with pointed ends. Longer scats often fold over upon themselves. Scats of fish or crayfish are typically tubular.

            Mink defecate on elevated surfaces, such as stumps (Gerell 1968), as well as bridges, logs, or grass clumps over water. Scats also accumulate at the entrances of den sites, as well as in the latrines of Muskrats and raccoons. Scats may accumulate at regular scent sites, near burrows used for denning or regular resting, as well as where territories of two or more mink overlap along watercourses (Brinck, Gerell, and Odham 1978).

URINE AND OTHER SCENT-MARKING BEHAVIORS: Mink have two anal scent glands on either side of the anus called anal pouches. They apply glandular scents by dragging their anal region along the ground. Mark Elbroch witnessed such a display in a wild mink marking a clump of grass along a watercourse. The mink spread its hind feet to flatten its hind end and place its anus in contact with the ground, and then rapidly dragged itself forward with its front limbs; the mink returned to the water after each pass, but then returned to re-mark the area four or five times.

    The anal pouches can also be forcefully evacuated and scent secretions shot up to 12 in. (30 cm) behind the animal. They don't squirt scent in the far-reaching, focused manner of skunks, but create a noxious cloud around themselves in times of stress (Brinck, Gerell, and Odham 1978).

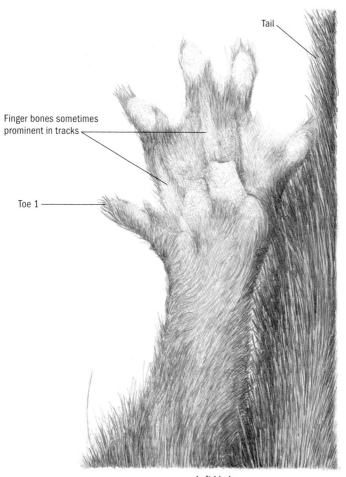

Tail

Finger bones sometimes
prominent in tracks

Toe 1

Left hind

American Mink foot.

Walking American Mink.

The 3× lope of an American Mink.

The 2 × 2 lope of an American Mink.

In certain substrates the toes and claws of mink appear very pointed, and the palm pads register only lightly if at all.

The loping track pattern of an American Mink. From bottom to top, the tracks are left front, overlaid with left hind slightly above, then right front and right hind.

AMERICAN BADGER

Taxidea taxus

Plate 24
Plates 111

TRACKS: Front: 2⁷/₈–3⁷/₈ in. long (7.3–9.8 cm) × 1⁹/₁₆–2⁵/₈ in. wide (4–6.7 cm)

Medium. Plantigrade. Asymmetrical. Five toes. Toe 1 is the smallest, located on the inside of the track, and does not register reliably. The metacarpal pads are fused and form a large pad, wider than long. The bottom of the foot is naked, and the foot rarely splays. Claws are large and long, and they register reliably; the claws of toes 2–4 register in very hard substrates, even when other parts of the tracks are obscured or impossible to see. The front tracks are significantly larger than the hind tracks.

Hind: 1⁷/₈–2¾ in. long (4.8–7 cm) × 1³/₈–2 in. wide (3.5–5.1 cm)

Small to medium. Plantigrade. Asymmetrical. Five toes. Toe 1 is smallest and often does not register. The metatarsals are fused to form one large pad; the posterior edge of the metatarsals is notched and is a key character in identifying badger tracks. Claws may or may not register. Hind tracks are much smaller than front tracks.

TRAIL: Walk: Strides: 5½–9¾ in. (14–24.8 cm)
Trail widths: 5–7½ in. (12.7–19.1 cm)

Trot: Strides: 10–15 in. (25.4–38.1 cm)
Trail widths: 3–6 in. (7.6–15.2 cm)

Lope: Strides: 10–12 in. (25.4–30.5 cm)
Group lengths: 17–20 in. (43.2–50.8 cm)

NOTES:

• Walk when foraging but often trot when traveling or crossing exposed areas. They lope when alarmed.

• American Badgers, as opposed to their European relatives, are creatures of grasslands, desert, and open country.

• In deep snow, badgers walk, creating a pigeon-toed trail in which their tracks are connected by contiguous drag marks.

• In cases where only four toes are visible in the hind track, badger tracks could easily be confused with those of Bobcats or similar-sized canids.

SCAT: ³/₈–1 in. (1–2.5 cm) diameter, 3–6 in. (7.6–15.2 cm) long

Badger scats vary in shape and dimensions considerably, but almost always release a strong odor when opened up to study their contents. Scats may be long and tubular, or very twisted, folded, pointed piles of somewhat segmented scats. If you are lucky, you will find them along travel routes or occasionally posted at the entrances of ground squirrel and prairie dog burrows. Badger scats are very difficult to find, because they typically defecate below ground. In 18 months of locating marked badgers in their burrows using telemetry, Quinn (2008) and helpers located only six scats.

URINE AND OTHER SCENT-MARKING BEHAVIORS: Badgers often rub their abdominal gland on the mound of soil outside their dens, leaving a mark that looks like they slid down the mound as they exited (Lindzey 2003). They also drag their abdominal glands in suitable substrate to scent-mark as they travel.

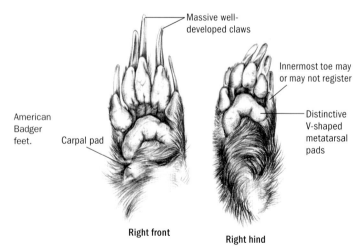

Massive well-developed claws

Innermost toe may or may not register

Distinctive V-shaped metatarsal pads

American Badger feet.

Carpal pad

Right front

Right hind

Trotting American Badger.   Walking American Badger.

The walking trail of an American Badger in sand dunes in the Mojave National Preserve.

American Badger tracks: larger left front above with long claws, and smaller left hind below, lacking the innermost toe 1.

An American Badger drags its stomach and hind end as it marks a soft section of earth at the edge of its territory.

Skunks: Family Mephitidae

STRIPED SKUNK *Mephitis mephitis*

TRACKS: Front: 1⁵/₈–2¹/₁₆ in. long (4.1–5.2 cm) × 1–1¹³/₁₆ in. wide (2.5–3 cm)

Small. Plantigrade. Asymmetrical. Five toes. Toe 1 is the smallest, located on the inside of the track. Toes are fused together, and the track cannot splay. The metacarpal pads form one larger pad. The carpal pad (the heel) may or may not register. The digital, palm, and heel pads are completely naked. The claws are large, long, and reliably register; in fact, on hard substrate, the claws from toes 2–4 may be all you can see.

Hind: 1⁵/₁₆–2 in. long (3.3–5.1 cm) × ¹⁵/₁₆–1¹³/₁₆ in. wide (2.4–3 cm)

Small. Plantigrade. Asymmetrical. Five toes. Toe 1 is the smallest and often does not register. Toes are partially fused (the central three toes are essentially one unit). The metatarsals are fused to form one large palm pad, separated by a large crease (a line in the footprint) from the large heel that more often registers than not. The digital, palm, and heel pads are naked. The claws on the hind feet are short and may or may not register.

TRAIL:

Direct-register walk:	Strides: 4–8 in. (10.2–20.3 cm)	
	Trail widths: 3–4½ in. (7.6–11.4 cm)	
Overstep walk:	Strides: 5¾–9 in. (14.6–22.9 cm)	
Direct-register trot:	Strides: 8–11 in. (20.3–27.9 cm)	
	Trail widths: 2³/₈–4½ in. (6–11.4 cm)	
Lope:	Strides: 3–7½ in. (7.6–19.1 cm)	
	Group lengths: 8–16½ in. (20.3–42 cm)	
Gallop:	Strides: 3¾–7¾ in. (9.5–19.7 cm)	
	Group lengths: 10–18 in. (25.4–45.7 cm)	

NOTES:

- Often explore and cover open areas in a rocking-horse lope, and slow down to investigate and forage in a walk.

- The trails of Striped Skunks are often found crossing wide open areas and marching straight down the middle of trails and roads.

- Differentiating trots and walks can usually be done by measurements, but also note that when trotting, any drag marks between tracks are straight, rather than curved, and that tracks are less pigeon-toed than when they are walking.

- In deep snow, Striped Skunks use a very pigeon-toed direct-register walk that can easily be confused with the trails of house cats.

• In deep substrates the toes do not splay, and the tracks are blocky and look like they were made by someone poking the end of a stick into the ground or mud.

SCATS: $^3/_8$–$^7/_8$ in. (1–2.2 cm) diameter, 2–5 in (5.1–12.7 cm) long

Skunk scats vary somewhat according to diet but are often tubular scats with smooth surfaces and blunt ends, which break easily when prodded. Striped Skunks sometimes drop scats at random, without even stopping to defecate, and spread the resulting scat over four or five steps. Striped Skunks also form latrines at den sites, under overhangs, elevated along travel routes, or near a prominent feature along travel routes, such as a stump in a field. Latrines are common in old sheds and abandoned buildings.

URINE AND OTHER SCENT-MARKING BEHAVIORS: Chemical communication is poorly understood in skunks, other than intensive research on the chemical compounds in their spray. Skunks are black-and-white animals, and this aposematic coloration no doubt communicates much to both conspecifics and potential competitors in a highly developed visual communication (Larivière and Messier 1996).

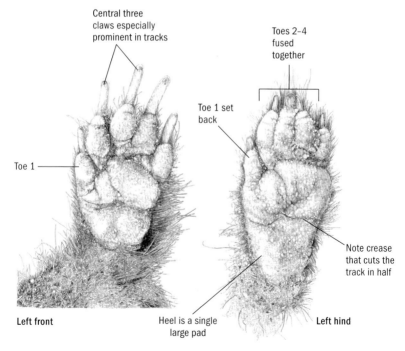

Central three claws especially prominent in tracks

Toes 2-4 fused together

Toe 1 set back

Toe 1

Note crease that cuts the track in half

Left front

Heel is a single large pad

Left hind

Striped Skunk feet.

Direct-register walk of
a Striped Skunk.

Overstep walk of a
Striped Skunk.

Loping Striped
Skunk.

Complete front and hind tracks of a Striped
Skunk.

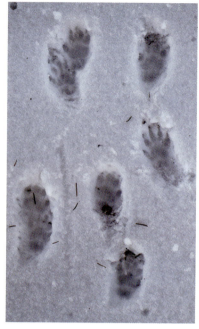

A cluster of Striped Skunk tracks in snow.

The latrine of a Striped Skunk under a rock overhang.

WESTERN SPOTTED SKUNK

Spilogale gracilis

EASTERN SPOTTED SKUNK

Spilogale putorius

Plate 4
Plate 16

TRACKS: Front: 1–1⁵/₈ in. long (2.5–4.1 cm) × ¾–1¹/₁₆ in. wide (1.9–2.7 cm)

Small. Plantigrade. Asymmetrical. Five toes. Toe 1 is the smallest, located on the inside of the track. Toes 2, 3, and 4 form a curved line, helping differentiate front tracks from hind ones. Unlike Striped Skunks, toes 2–4 are only partially fused, allowing the toes some independent movement. The metacarpal pads also register more independently than in tracks of Striped Skunks. Two additional pads (the heel) register to the posterior, the larger of which is the carpal pad, which registers more reliably than the other. All the digital, palm, and heel pads are naked. Claws are larger and longer than those found in hind tracks, but still significantly shorter than the claws of the front feet of Striped Skunks. Front and hind tracks are of similar dimensions.

Hind: ⁷/₈–1³/₈ in. long (2.2–3.5 cm) × ¾–1 in. wide (1.9–2.5 cm)

Small. Plantigrade. Asymmetrical. Five toes. Toe 1 is smallest, and distinctly farther back in the track than toe 5. Toes 2, 3, and 4 form a straight line and register as a unit, helping differentiate front from hind feet. The metatarsals are fused to form one large pad. Two separate heel pads often register, as opposed to the single large heel pad seen in tracks of Striped Skunks. The digital, palm, and heel pads are naked. Claws may or may not register.

TRAIL: Walk: Strides: 2–4 in. (5.1–10.2 cm)

Overstep walk: Strides: 5½–6½ in. (14–16.5 cm)
Trail widths: 3–4 in. (7.6–10.2 cm)

Bound: Strides: 6–30 in. (15.2–76.2 cm)
Trail widths: 2¼–4 in. (5.7–10.2 cm)

Lope: Group lengths: 9–11 in. (22.9–27.9 cm)

NOTES:

- Travel in lopes and bounds, and slow to walking gaits to explore and forage. At first glance, Spotted Skunk bound patterns can be mistaken for those of a squirrel.

- Can climb trees, and often rest in tree cavities or crevices in rock walls.

SCAT: ¼–¾ in. (0.6–1.9 cm) diameter

Skunk scats are often tubular scats with smooth surfaces and blunt ends, which break easily when prodded.

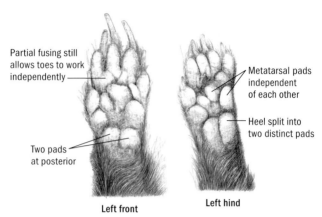

Partial fusing still
allows toes to work
independently

Metatarsal pads
independent
of each other

Heel split into
two distinct pads

Two pads
at posterior

Left front

Left hind

Western Spotted Skunk feet.

Bounding spotted skunk.

Light, but perfect tracks of a Western Spotted Skunk. Study the front right track on the left, and note the paired pads at the posterior of the footprint. The right hind track on the right also exhibits paired pads at the posterior of the footprint.

Spotted skunk tracks circumventing a rock along a ctroam bank.

Ringtail and Raccoon: Family Procyonidae

RINGTAIL (RING-TAILED CAT) *Bassariscus astutus*

TRACKS: Front: 1⅛–1⁷/₁₆ in. long (2.9–3.7 cm) × ¹⁵/₁₆–1¼ in. wide (2.4–3.2 cm)

Small. Plantigrade. Symmetrical. Five toes. Toe 1 is the smallest, located on the inside of the track, and does not register reliably. The metacarpal pads are fused to form a larger pad. The carpal pad (the heel) is large and registers more often than not. Digital, palm, and heel pads are bulbous, soft, and naked. Claws are semiprotractible and do not register in most tracks.

Hind: 1–1³/₈ in. long (2.5–3.5 cm) × ⁷/₈–1¹/₁₆ in. wide (2.2–2.7 cm)

Small. Plantigrade. Asymmetrical. Five toes. Toe 1 is smallest and rarely registers well in tracks when they are walking. The metatarsals are fused to form one large pad, much longer than wide. The digital, palm, and heel pads are naked. Claws are semiprotractible and do not register in most tracks. Because toe 1 is often absent, the hind track can be confused with a very small Domestic Cat track; however, note the unusually large palm pad, which takes up a large portion of the track.

TRAIL: Direct-register walk: Strides: 3–8 in. (7.6–20.3 cm)
Trail widths: 1¼–3½ in. (3.2–8.9 cm)

Bound: Strides: 12–48 in. (30.5–121.9 cm)
Trail widths: 3–4½ in. (7.6–11.4 cm)

NOTES:

• Walk in their explorations and hunting forays, but will speed up into bounds and lopes when they feel exposed, threatened, or are chasing prey. One of the few species that regularly uses an understep walk.

• Tend to stay near cover, whether that be brush, trees, or rocky cliffs. They are especially common along the large riparian corridors in northern California, including along the Sacramento River north of Sacramento.

• Tend to be solitary animals, but paired trails may be encountered throughout the year.

SCAT: ³/₁₆–⁵/₈ in. (0.5–1.6 cm) diameter, 2½–5 in. (6.4–12.7 cm) long

Ringtail scats are highly variable, depending upon diet. When composed of fur and bone, they are twisted, tapered ropes that resemble those made by weasels. Scats are generally found elevated above a travel route on stumps, rocks, or large boulders, or in more suburban settings, placed along the margins of trails and roads. In a study near

Mexico City, 62 of 80 defecation sites held more than one scat, and some latrines held as many as 19 (Barja and List 2006).

URINE AND OTHER SCENT-MARKING BEHAVIORS: Urine is frequently rubbed on prominent objects, such as stumps and rocks.

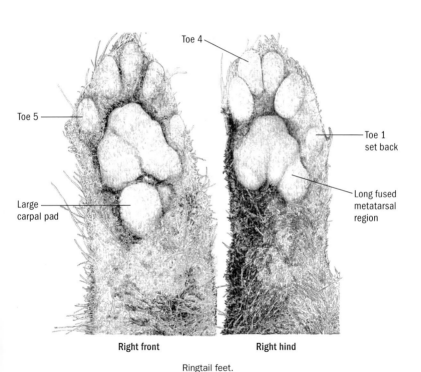

Toe 4

Toe 5

Toe 1
set back

Large
carpal pad

Long fused
metatarsal
region

Right front   Right hind

Ringtail feet.

Bounding to loping
Ringtail.

Walking Ringtail.

Clean Ringtail tracks in a fine layer of dust: the left hind track on the left lacks the innermost toe, but the left front track on the right has registered completely.

Pristine tracks of a Ringtail in a silt wash along the Sacramento River after a flooding event. Both tracks are the left side of body, the front above includes the carpal pad and the hind sits below; look closely for the marks made by the claws.

NORTHERN RACCOON   *Procyon lotor*

Plate 10
Plate 28
Plates 112 and 113

TRACKS: Front: 1¾–3⅛ in. long (4.4–7.9 cm) × 1½–3¼ in. wide (3.8–8.3 cm)

Medium. Plantigrade. Symmetrical. Five toes. The digital pads often appear cigar-shaped and connect to the metacarpal pads. The metacarpal pads are fused to form a larger pad, which appears distinctly curved, or C-shaped, in tracks. The carpal pad (the heel) occasionally registers to the posterior. The digital, palm, and heel pads are naked. Claws may or may not register. Tracks may resemble miniature human handprints.

Hind: 2⅛–3⅞ in. long (5.4–10.2 cm) × 1½–2⅝ in. wide (3.8–6.7 cm)

Medium. Plantigrade. Asymmetrical. Five toes. Toe 1 is the smallest toe and is situated behind toes 2–5. The digital pads often connect to the metatarsal pads. The metatarsals are fused to form one large pad, much longer than wide. The complete, triangular back portion of the heel may or may not register. The digital, palm, and heel pads are naked. Claws may or may not register in tracks. When toe 1 does not register (which is common in shallow substrates), the hind track can be confused with a Bobcat track.

TRAIL:

Walk 2 × 2:	Strides: 8–19 in. (20.3–48.3 cm)	
	Trail widths: 3½–7 in. (8.9–17.8 cm)	
Direct-register walk:	Strides: 8–14 in. (20.3–35.6 cm)	
	Trail widths: 3½–6 in. (8.9–15.2 cm) trots down to 2½ in. (6.4 cm)	
Pacelike walk (fast):	Strides: 18–25 in. (45.7–63.5 cm)	
	Trail widths: 3½–6 in. (8.9–15.2 cm)	
Bound:	Strides: 10–30 in. (25.4–76.2 cm)	
	Group lengths: 15–27 in. (38.1–68.6 cm)	

NOTES:

- Use a distinctive 2 × 2 walk in their travels and foraging. Lopes and gallops are used when alarmed or threatened.

- Trails often lead straight from dens or holes to good feeding areas, with little wandering done in between; trails can be quite long and straight.

- Will explore and use abandoned burrows, culverts, basements, and other available hollows.

- In deep snow, raccoons use a direct-register walk, which can be confused with a walking Fisher or even a Bobcat. Typically their tracks are narrow, however, because they tend not to splay their toes in deep snow.

SCAT: ⁵/₁₆–1³/₁₆ in. (0.8–3 cm) diameter, 3½–7 in. (8.9–17.8 cm) long

Warning: raccoon scats are known to carry the parasitic roundworm, *Baylisascaris procyonis,* which in humans can cause illness or death. Handle with caution.

The diets of raccoons are highly variable, ranging from fruits and mast to scavenging in garbage cans to foraging for mole crabs and crayfish, and this variety is reflected in the diverse forms of scats they produce. Classic raccoon scats are tubular, break easily, fold back on themselves, and have blunt ends; they are similar to bear scats, but smaller. However, they also produce amorphous blobs, patties, and occasionally twisted scats as well. Regardless of diet, raccoon scats hold a very earthy, distinctive scent. It's a deep, musty odor, not unpleasant but very strong. Many people are reluctant to smell raccoon scats, given that there is the tiniest possibility that the eggs of *Baylisascaris procyonis* could be inhaled. Certainly do not attempt to smell dry raccoon scats, but when you encounter moist scats of questionable origin, smell can be a great clue in being more certain you've found the scat of a raccoon.

Raccoons form large latrines at the bases of dominant conifers, especially those near wetland systems and water sources; latrines are most often on the side facing away from the water source. They also form large latrines under rock overhangs, and in other human structures that mimic such circumstances, such as attics or abandoned buildings. Raccoons defecate more frequently in latrines during summer, fall, and winter, and deposit scats more randomly during the spring (Stains 1956).

URINE AND OTHER SCENT-MARKING BEHAVIORS: Ough (1982) studied a group of captive raccoons sharing an enclosure and found that individuals maintained their own personal scent posts with neck and anus rubs, but formed communal latrines of scat.

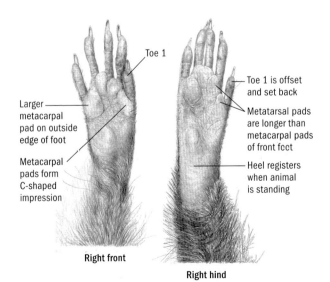

Toe 1

Toe 1 is offset and set back

Metatarsal pads are longer than metacarpal pads of front feet

Larger metacarpal pad on outside edge of foot

Metacarpal pads form C-shaped impression

Heel registers when animal is standing

Right front

Right hind

Northern Raccoon feet.

Bounding into galloping Northern Raccoon.

The 2 × 2 walk of a Northern Raccoon.

Fast 2 × 2 walk of a Northern Raccoon.

Slow 2 × 2 walk of a Northern Raccoon.

Gorgeous front (right side of photo) and hind tracks of a Northern Raccoon under a bridge in Santa Barbara, California, surrounded in roof rat tracks.

The 2 × 2 pattern created by a rapidly walking Northern Raccoon, where the angle of lines made through the middle of each pair of tracks alternates right then left.

A Northern Raccoon's latrine of several earthy, soft scats along the Sacramento River.

Northern Raccoon latrine under overhanging boulders.

ORDER PERISSODACTYLAD

Feral Horse: Family Equidae

Plate 33

Plates 100 and 125

ASS (WILD BURRO, AFRICAN WILD ASS)   *Equus asinus*

TRACKS: Front: 3¼–3¾ in. long (8.2–9.4 cm) × 2½–2⅝ in. wide (6.3–6.7 cm)

Hind: 2⅝–3⅜ in. long (6.7–8.5 cm) × 2¼–2½ in. wide (5.7–6.3 cm)

Data compliments of Barry Martin and Casey McFarland, who collected these parameters from three Wild Burros in eastern San Diego County.

For a description of Wild Burro tracks, see Wild Horse, in the next account.

WILD HORSE   *Equus caballus*

TRACKS: Front: 4½–5½ in. long (11.4–14 cm) × 4–5¼ in. wide (10.2–13.3 cm)

Hind: 4¼–5¼ in. long (10.8–13.3 cm) × 3¾–4¾ in. wide (9.5–12.1 cm)

Large. Symmetrical. Three toes. The remnants of toes 2 and 4 are raised up on the backside of the leg and appear as oval calluses, sometimes referred to as chestnuts, that do not register in the track. Toe 3 forms the entire round hoof; the posterior edge has two rounded lobes separated by a notch. This feature is often referred to as the "frog" of the track and somewhat resembles a deer track. The front track is larger and proportionally wider than the hind track.

TRAIL: Walk: Strides: 23–38 in. (58.4–96.5 cm)

                    Trail widths: 7–16 in. (17.8–40.6 cm)

NOTES:

· Walk about when foraging, but play using lopes and gallops.

· Horseshoes are often apparent in tracks. Trained horses may use a variety of gaits not seen in wild animals.

· Feral horses in California are primarily descendants of Spanish stock escaped after World War I.

SCAT: Pellets 1–1¾ in. (2.5–4.4 cm) diameter, 2–3 in. (5.1–7.6 cm) long

Horse droppings are often a larger clump of smaller pelletlike defecations. Pellets are large (at least three or four times as large as elk pellets), longer than wide, and roughly shaped like a kidney bean. Typical color variations range from green to brown, and darken with age. Horse scats can very easily be confused with scats of bears when bears are feeding on vegetation; however, horses chew vegetation much more thoroughly, and the defecated remains will be much finer than the coarse, longer pieces of vegetation in bear scats. Bears also eat numerous stringy roots, which horses generally do not, so breaking up questionable scats will reveal the size of the individual pieces of vegetation and whether they are root, grass, or hay, which will aid in species identification.

Wild Burro track, photographed near Palm Springs, California. Photo by Barry Martin.

Walking feral horse.

Wild Horse tracks.

ORDER ARTIODACTYLA

Wild Boar: Family Suidae

WILD BOAR (FERAL HOG) *Sus scrofa*

Plate 27
Plates 117 and 121

TRACKS: Front: $2^1/_8$–$2^5/_8$ in. long (5.4–6.7 cm) × 2¼–3 in. wide (5.7–7.6 cm)

Hind: $1^7/_8$–2½ in. long (4.8–6.4 cm) × 2–2¾ in. wide (5.1–7 cm)

Medium. Slightly asymmetrical. Four toes. Toes 2 and 5 are dew-claws, situated above toes 3 and 4 on the backside of the leg. Toes 3 and 4 register in tracks; toe 3 is slightly smaller than toe 4. Dewclaws are common in tracks, and register just behind and to the outside of the cleaves. Hooves are blunt and rounded, although young may be pointed; inner walls slightly concave. Front larger than hind track. Dewclaws are not included in measurements.

TRAIL: Walk: Strides: 13–17 in. (33–43.2 cm)
Trail widths: 7–8½ in. (17.8–21.6 cm)

Direct-register trot: Strides: 26–32 in. (66–81.3 cm)
Trail widths: 3–4½ in. (7.6–11.4 cm)

Lope/gallop: Strides: 36–52 in. (91.4–132.1 cm)
Group lengths: 46–52 in. (116.8–132.1 cm)

NOTES:

- The feral pigs in California are hybrid animals—descendants of escaped domestic stocks introduced to the state since as early as the 1700s and wild European Wild Boars, introduced in the 1920s. In behavior they follow closely with European Wild Boars, but in appearance they are a mix of both domestic and feral strains. Scattered populations still persist throughout California, despite concerted efforts to remove them.

- Walk about the landscape, feeding on the move. They trot to cover ground more quickly, and bound and gallop when alarmed.

- Look for obvious signs of rooting wherever hogs are moving.

- May be found moving alone or in large, social groups.

- The width of the dewclaws is a useful feature for identification. In Wild Boars, the dewclaws are often wider than the track width. In deer tracks, the dewclaws are found directly behind the cleaves.

SCAT: 1–2 in. (2.5–5.1 cm) diameter, 4–14 in. (10.2–35.6 cm) long

Wild pig scat is highly variable, depending upon the water content in their diet. When they are eating vegetation high in water content, their scats are soft, irregular lumps longer than wide. When they are feeding on mast crops or fruits, or scavenging garbage or carcasses, their scats are tubular and segmented, and can very easily be confused with those of black bears.

URINE AND OTHER SCENT-MARKING BEHAVIORS: Feral pigs use at least three glands in scent marking: (1) They rub tusk glands (found in the upper lip above the tusk) on trees and stumps, where they may also inadvertently score the bark in the process. Tusk sites may also be used for body scratching, where coarse hair or mud can be left behind as a visible clue. (2) Males, called boars, produce a strong-smelling urine when using the preputial gland in their urethra. (3) A boar may lean forward and scrape at the ground with its feet to mark with its metacarpal glands, located on the posterior portion of the front feet (Sweeney, Sweeney, and Sweeney 2003).

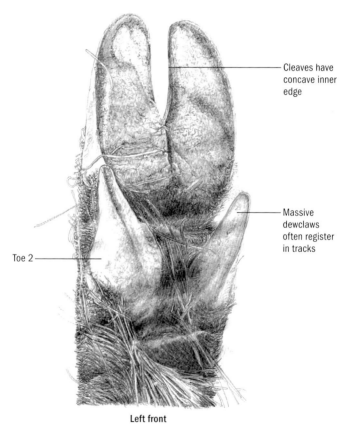

Cleaves have concave inner edge

Massive dewclaws often register in tracks

Toe 2

Left front

Wild Boar foot.

Trotting Wild Boar.

Walking Wild Boar.

Wild Boar tracks in shallow substrate do not show the dewclaws. Photo by Jonah Evans.

Compare the walking trail of a Wild Boar on the left with the walking trail of a Mule Deer on the right. Coyotes, too, are cruising down the middle of the photo.

The wallow and rubbing post of a Wild Boar near Santa Cruz, California.

Deer and Relatives: Family Cervidae

MULE DEER	***Odocoileus hemionus***
Subspecies: BLACK-TAILED DEER	***O. h. columbianus***

Plate 25
Plates 99 and 120

TRACKS: Front: 2¼–4 in. long (5.7–10.2 cm) × 1⅝–2¾ in. wide (4.1–7 cm)

Hind: 2– 3½ in. long (5.1–8.9 cm) × 1½–2⅜ in. wide (3.8–6 cm)

Medium. Unguligrade. Slightly asymmetrical. Four toes. Toes 2 and 5 are dewclaws, situated above toes 3 and 4 on the backside of the leg (the dewclaws are situated higher on the hind legs than the front). Toes 3 and 4 register in tracks; toe 3 is usually slightly smaller than toe 4. Hooves are heart-shaped, pointing in the direction of travel; interior hoof walls are concave for the forward half of the track. Soft pads fill tracks almost completely, and the subunguinis is a thin strip between the pads and hoof walls. Front tracks larger than hind tracks. Tracks splay widely, each cleave angling out as much as 40 degrees from the midline, due to substrate and/or speed.

TRAIL:	Walk:	Strides: 15–25½ in. (38.1–64.8 cm)
		Trail widths: 5–10 in. (12.7–25.4 cm)
	Straddle trot:	Strides: 33–50 in. (83.8–127 cm)
		Group lengths: 7–16 in. (17.8–40.6 cm)
	Pronk:	Strides: 8–15 ft (243.8–457.2 cm), and occasionally longer.
		Group length: 30–42 in. (76.2–106.7 cm)

NOTES:

• When moving quickly or in deep substrates, dewclaws are large and obvious. The dewclaws of the front tracks point laterally, while those of the hind tracks point in line with the direction of the track. As opposed to the dewclaws of feral hogs, the dewclaws of deer register directly behind the cleaves.

• Walk when traveling or foraging, and then speed up into trots, lopes, and gallops when alarmed or threatened.

• Walk in snow, dragging their feet from one step to the next, even in shallow snow; as snow deepens, they form well-used runs.

• "Point" when browsing, meaning a single front footprint will angle off from the trail and point in the direction of the browsed vegetation.

SCAT: Pellets: ³/₁₆–⅝ in. (0.5–1.6 cm) diameter, ½–1¾ in. (1.3–4.4 cm) long

Deer scats vary with the moisture content in their diet, which is dependent upon seasonal resources and weather. Wet-season scats are loose and amorphous as a result of the high moisture content of herbaceous plants that they feed upon at that time. As diets shift into drier vegetation, pellets become obvious within scats, but they remain soft and clumped together. When defecating pellets, members of the deer family expel a group of pellets simultaneously, as opposed to rabbits, which excrete one at a time. During the dry season, a woody diet produces harder pellets that hold their shape well and persist for a long time on the landscape. Often there is a dimple at one end and a point at the other, but they may be rounded at both ends as well. Pellets may be short and stout, or long and skinny.

Wigley and Johnson (1981) studied the disappearance rates of deer pellets and found that they were unrecognizable within a month in the wet season of the southeastern United States, but were identifiable for well over a year in dry areas in the west.

URINE AND OTHER SCENT-MARKING BEHAVIORS: Urine often smells piney, light, and somewhat pleasant.

In the rut, Mule Deer bucks grind the bases of their antlers and foreheads on small trees to mark them with scent, and thrash vegetation with their antlers. By raking the antlers back and forth in shrubs and branches, they create loud auditory dominance displays.

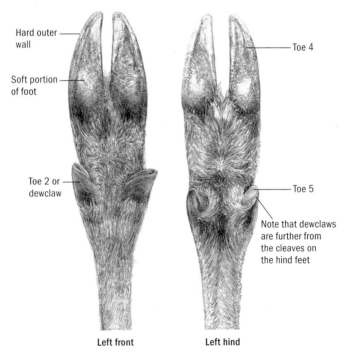

Mule Deer feet.

Stotting Mule
Deer.

Walking Mule
Deer.

Trotting Mule
Deer.

Front and hind tracks of a Mule Deer.

Tracks of the Black-tailed subspecies of Mule Deer: the front track is above and the hind below.

ELK (WAPITI)

Cervus elaphus

Plate 32
Plates 99, 121, and 122

TRACKS: Front: 3–4⁷/₈ in. long (7.6–12.4 cm) × 2⁵/₈–4⁵/₈ in. wide (6.7–11.7 cm)

Hind: 2½–4½ in. long (6.4–11.4 cm) × 2³/₈–4 in. wide (6–10.2 cm)

Medium to large. Unguligrade. Slightly asymmetrical. Four toes. Toes 2 and 5 are dewclaws, situated above toes 3 and 4 on the backside of the leg. Toes 3 and 4 register in tracks; toe 3 is usually slightly smaller than toe 4. Hooves rounded, sometimes pointed with wider tips; interior hoof walls slightly concave. In front tracks, the soft pads at the posterior edge form approximately one-quarter of the overall track length; the subunguinis fills the remainder of tracks. In hind tracks, the soft pads at the posterior edge form approximately one-third of the overall track length. The front track is larger than the hind track. Cleaves can spread and splay due to substrate and/or speed, though not to the extent of Mule Deer.

TRAIL:	Walk:	Strides: 18–35 in. (45.7–88.9 cm)
		Trail widths: 6¾–13 in. (17.1–33 cm)
	Trot:	Strides: 32–45 in. (81.3–114.3 cm)
		Trail widths: 3½–6 in. (8.9–15.2 cm)
	Straddle trot:	Strides: 37–49 in. (94–124.5 cm)

NOTES:

- Walk when traveling or foraging, and then speed up into trots, lopes, and gallops when alarmed or threatened.
- Direct-register walk in snow, leaving contiguous drag marks between tracks.

SCAT: Pellets: ⁷/₁₆–¹¹/₁₆ in. (1.1–1.7 cm) diameter, ½–1 in. (1.3–2.5 cm) long

Elk scats vary with the moisture content in their diet, which is dependent upon seasonal resources and weather. Scats found during the wet season are loose and amorphous, as would be expected with the high moisture content of herbaceous plants that they feed upon at that time. As drier vegetation becomes primary in the diet, pellets become obvious within scats, but they remain soft and clumped together. During the dry season, a woody diet produces harder pellets that hold their shape well and persist for a long time on the landscape. An amorphous patty and hard pellets are extremes on a spectrum of possible variations, so remain flexible in your interpretations.

URINE AND OTHER SCENT-MARKING BEHAVIORS: An Elk begins making a scent post by sniffing a tree about 3.25 ft (1 m) off the ground, and then peeling or scraping off the bark with its incisors or antlers. The animals alternate licking and peeling, but don't eat the bark. Then they rub their foreheads, preorbital glands, necks, and shoulders on the exposed wood. Both sexes and all ages of Elk create scent posts, and they are made throughout the year.

"Horning" and "thrashing" are terms that refer to vigorously raking the antlers through low vegetation and against small trees. The antlers break up branches, strip off bark, and tear up turf and vegetation. One of the most frequent behaviors of rutting bulls is the thrash-urination. They thrash their antlers side to side through low vegetation, dig into the ground, and throw bits of earth and plant material up into the air. All the while, the penis is unsheathed and swinging, soaking the belly and neck, and surrounding ground with urine.

Thrash-urinations are the most frequent scenting display during the rut, and it is linked to direct aggression between bulls. In one study, dominant Roosevelt Elk bulls tending a harem thrash-urinated more than once every hour (Bowyer and Kitchen 1987). Bachelor bulls averaged only once every four hours, and yearlings were not observed exhibiting this behavior at all.

Wallows are made by rutting bulls on damp, soft ground, in much the same way they thrash-urinate. A bare patch of earth is exposed by thrashing and digging with the antlers and pawing with the front hooves. As the wallow is being dug, the bull spray-urinates in the pit. Once the pit is dug, the bull lowers himself, and rolls and rubs his mane in it, caking himself in his odors. Bulls make several wallows and re-use them frequently. Wallowing is somewhat rare in Tule Elk.

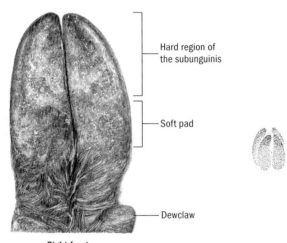

Hard region of
the subunguinis

Soft pad

Dewclaw

Right front

Elk foot.

Walking Elk.

Elk tracks, the hind landing on and overlapping with the larger front track.

A bull Elk antler rub.

Pronghorn: Family Antilocapridae

PRONGHORN *Antilocapra americana*

Plate 26

Plates 98 and 120

TRACKS: Front: 2¹/₈–3½ in. long (5.4–8.9 cm) × 1½–2¼ in. wide (3.8–5.7 cm)

Hind: 2¼–3¼ in. long (5.7–8.3 cm) × 1½–2¹/₈ in. wide (3.8–5.4 cm)

Medium. Unguligrade. Slightly asymmetrical. Two toes. Pronghorn lack dewclaws, and toes 3 and 4 register in tracks; toe 3 is slightly smaller than toe 4. Hooves heart-shaped; interior hoof walls concave for the forward half of the track, and outside walls either concave at about the center point, or straight. The soft pads at the posterior edge of tracks are distinctly bulbous. Front tracks are very similar dimensions to hind tracks, making them at times difficult to distinguish; splay often due to substrate and/or speed.

TRAIL: Direct-register walk: Strides: 17–26 in. (43.2–66 cm)
                              Trail widths: 4³/₈–10 in. (11.1–25.4 cm)

Overstep walk: Strides: 19–29 in. (48.3–73.7 cm)

Bound/gallop: Strides: 29–61 in. (73.7–154.9 cm)
                              Group lengths: 36–82 in. (91.4–208.3 cm)

NOTES:

• Overstep walk (unique among California ungulates and a useful feature for identification) or direct-register walk when traveling or foraging.

• They speed up into trots, lopes, and gallops when alarmed or threatened.

• Inhabit open grasslands and sagebrush deserts, relying upon speed and sight to keep them safe while so exposed. They can run at speeds up to 60 miles per hour (97 km per hour).

• Direct-register walk in deep snow.

SCAT: Pellets: ³/₁₆–½ in. (0.5–1.3 cm) diameter, ³/₈–¾ (1–1.9 cm) long

Pronghorn scats, much like those of deer, vary with diet and moisture content, both of which are dependent upon resource availability and weather. Wet-season scats are looser due to the high moisture content of herbaceous plants. As diets shift to drier vegetation, pellets become obvious within scats, but they are soft and clumped together. During the dry months, a woody diet produces harder pellets, which hold their shape well and last long into summer and sometimes beyond.

Strings of connected Pronghorn pellets—20 or so per string—may be encountered. Deer are not known to produce such sign.

URINE AND OTHER SCENT-MARKING BEHAVIORS: Dominant bucks mark bushes and other tall vegetation in their territories with a secretion from the subauricular gland located just anterior to and below their eye (Kitchen 1974). They also mark territories with urine and scats in a ritualistic pattern: in sequence, they use stiff forelimbs to scratch the earth, then they urinate on or adjacent the scrape, and finally they defecate (sometimes called a "SPUD"). Bachelor males mark in the same way, but outside defended territories (Byers 2003).

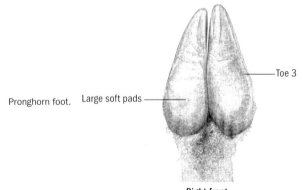

Pronghorn foot.   Large soft pads ———

———Toe 3

Right front

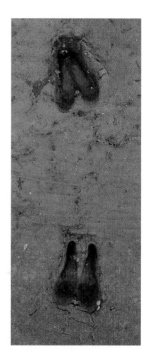

Front and hind tracks of a Pronghorn, exhibiting the concave outer edges to their heels.

Direct-register walk of a Pronghorn.

Overstep walk of a Pronghorn.

A male Pronghorn's SPUD.

Cattle, Bison, Goats, and Sheep: Family Bovidae

AMERICAN BISON *Bison bison*

Plate 34

Plates 125 and 126

TRACKS: Front: 4½–6½ in. long (11.4–16.5 cm) × 4½–6 in. wide (11.4–15.2 cm)

Hind: 4¼–6 in. long (10.8–15.2 cm) × 4–5½ in. wide (10.2–14 cm)

Large to very large. Unguligrade. Slightly asymmetrical. Four toes. Toes 2 and 5 are dewclaws and are situated above toes 3 and 4 on the backside of the leg. Toes 3 and 4 register in tracks; toe 3 is slightly smaller than toe 4. Tracks are round—front tracks more so than hind ones; the interior hoof walls are concave for the entire length of the track, and the negative space between the cleaves is large mid-track. Hind hooves are also very rounded; the interior hoof walls are slightly concave for the entire length of the track, but the negative space between the cleaves is very small for the entire length of the track. The soft pads in both front and hind tracks are distinctly V-shaped at the posterior edge. Front track is larger than hind track. Tracks may splay due to substrate and/or speed. Tracks are very similar and easily confused with those of domestic cows.

TRAIL: Walk: Strides: 22–38 in. (55.9–96.5 cm)
Trail widths: 10–22 in. (25.4–55.9 cm)

Trot: Strides: 40–44 in. (101.6–111.8 cm)
Trail widths: 10–15 in. (25.4–38.1 cm)

NOTES:

- Tracks of mature bulls are larger than females.
- Walk when traveling or foraging. They speed up when alarmed, threatened, or charging intruders.
- In California, bison are farmed for meat and other products; they are notorious for breaking out of fenced enclosures.
- Plow through deep snow, and you'll often find trails where they swing their head from side to side to dig down to grasses and other vegetation.
- Tend to be found in groups, but solitary animals are not uncommon.

SCAT: Patties: 10–16 in. (25.4–40.6 cm) diameter

Chips: Diameter of individual chips ranges from 3 to 4½ in. (7.6–11.4 cm)

Cows and bison shift from amorphous, flat patties in the wet season when they feed on herbaceous vegetation, to "chips" in the dry season, which look like smaller layered patties that hold their form. The patties and chips of bison and domestic cows are nearly indistinguishable from each other.

URINE AND OTHER SCENT-MARKING BEHAVIORS: Like elk, bison bulls often urinate in their dust baths and wallows before and during the rut. They may also urinate on themselves and then roll. Scent in the urine can communicate identity, age, status, or reproductive condition (McMillan et al. 2000). Horning and rubbing also probably deposit scent that conveys information to other bison (Coppedge and Shaw 1997).

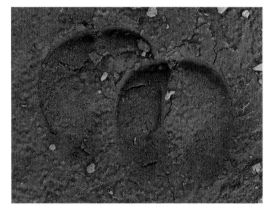

Walking American Bison.   Front and hind tracks of an American Bison.

The well-worn dust bath of an American Bison.

BIGHORN SHEEP *Ovis canadensis*

Plate 26

Plates 99 and 121

TRACKS: Front: 2¹/₈–3³/₈ in. long (5.4–8.6 cm) × 1½–3 in. wide (3.8–7.6 cm)

Hind: 2¹/₁₆–3¼ in. long (5.2–8.3 cm) × 1½–2³/₈ in. wide (3.8–6 cm)

Medium. Unguligrade. Slightly asymmetrical. Four toes. Toes 2 and 5 are dewclaws, situated above toes 3 and 4 on the backside of the leg. Toes 3 and 4 register in tracks; toe 3 is slightly smaller than toe 4. Hooves may be pointed and somewhat heart-shaped or blocky with blunt tips. Front track is larger than hind track.

TRAIL: Walk: Strides: 12–25 in. (30.5–63.5 cm)
Trail widths: 4–11 in. (10.2–27.9 cm)

Trot: Strides: 23–35 in. (58.4–88.9 cm)
Trail widths: 3–6 in. (7.6–15.2 cm)

Lope/gallop: Group lengths 50–60 in. (127–152.4 cm)

NOTES:

• Walk in steep and rocky terrain, but more often trot when traveling on flat ground or feeling exposed.

• Direct-register walk in deep snows.

SCAT: Pellets: ¼–⅝ in. (0.6–1.6 cm) diameter, ⅜–1 in. (1–2.5 cm) long

Bighorn Sheep scats vary with the moisture content in their diet, which is dependent upon seasonal resources and weather. Loose and amorphous scats are a result of the high moisture content of herbaceous plants that they feed upon in the wet season. With the start of the dry season, diets shift into drier vegetation; pellets become obvious within scats, but they remain soft and clumped together. A woody diet is consumed during the colder months, producing harder pellets that hold their shape well and last long into summer and sometimes beyond.

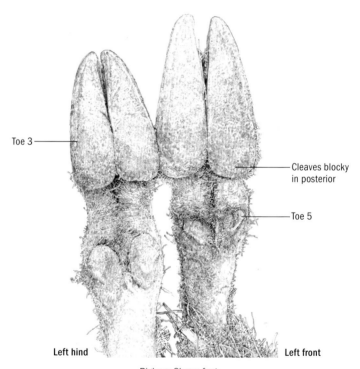

Bighorn Sheep feet.

Bighorn Sheep tracks in snow, front below and hind track above. Photo by Jonah Evans.

Walking Bighorn Sheep.

Bighorn Sheep tracks in mud.

ORDER RODENTIA

Mountain Beaver: Family Aplondontidae

MOUNTAIN BEAVER (SEWELLEL)   *Aplodontia rufa*

Plate 16

TRACKS: Front: $^{15}/_{16}$–1$^9/_{16}$ in. long (2.4–4 cm) × 1–1$^9/_{16}$ in. wide (2.5–4 cm)

Small. Plantigrade. Asymmetrical. Five toes. Toe 1 is reduced, lacks a claw, and rarely registers in tracks. The metacarpal pads are fused and lobed; there are two additional pads (one of which is the carpal pad) that form the "heel" at the posterior edge of the track, and they may or may not register. The negative space between the toes and metacarpals has little to no fur. Claws are long and prominent in tracks. Front and hind tracks are similar dimensions. The outline of the front track is shaped much like a miniature human hand.

Hind: $^7/_8$–1$^5/_8$ in. long (2.2–4.1 cm) × $^{15}/_{16}$–1½ in. wide (2.4–3.8 cm)

Small. Plantigrade. Slightly asymmetrical. Five toes. Toe 1 is smallest and often appears sharply pointed in tracks. The hind track is a classic rodent hind foot: toes 1 and 5 point out to either side, and toes 2, 3, and 4 register close together, pointing forward. The metatarsals are fused to form a larger pad. Claws may or may not register, and are much smaller than those of the front feet. The heel portion of the foot may or may not register.

TRAIL: Walk: Strides: 3¾–7 in. (9.5–17.8 cm)
                          Trail widths: 3–6¾ in. (7.6–17.1 cm)

NOTES:

• Understep-walk when traveling and foraging, but bound and gallop when alarmed.

• Tend to live under significant cover in moister habitats in northern California and riparian areas in portions of the Sierra Nevadas.

• Competent climbers, spend a great deal of time harvesting cambium, buds, and leaves from trees and shrubs.

• Follow a trail any distance, and it will inevitably start and end at a burrow entrance.

SCAT: $^3/_{16}$ in. (0.5 cm) diameter; ½ in. (1.3 cm) long

Mountain Beaver scat is cylindrical and small, with rounded ends. Their feces are very difficult to find, because they defecate below ground. Mountain Beavers practice coprophagy, meaning they eat their feces for a second run in the digestive tract to maximize energy extraction from a given meal. Soft pellets are produced first. Pellets produced by feces ingestion are smaller and hard. They are deposited directly into latrine compartments, or the animal tosses them into latrines with its mouth (Carraway and Verts 1993).

URINE AND OTHER SCENT-MARKING BEHAVIORS: Mountain Beavers regularly rub secretions in and around their nest chamber, and also scent-mark in outward tunnels at the periphery of their network used by neighboring animals. Other Mountain Beavers that encounter these scent marks spend a great deal of time investigating them but do not retreat, so researchers assume they do not serve as intraspecific barriers (Nolte et al. 1993).

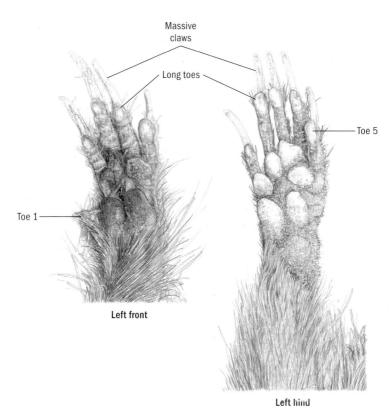

Mountain Beaver feet.

Walking Mountain Beaver.

The walking trail of a Mountain Beaver.

Squirrels: Family Sciuridae

LARGE TREE SQUIRRELS *Sciurus* spp.

Plate 4

Plates 17 and 18

Plate 96

TRACKS: Front: 1¼–2 in. long (3.2–5.1 cm) × ⅝–1⅜ in. wide (1.6–3.5 cm)

Small. Plantigrade. Asymmetrical. Five toes in classic rodent structure: toe 1 is a vestigial thumb, and unlike the toe 1 of the Douglas's Squirrel, rarely registers in tracks. Toes 2 and 5 point forward and to the sides, and toes 3 and 4 point forward. Some of the metacarpal pads are fused, but the front tracks include three distinct palm pads; there are two additional pads—the metacarpal pad for toe 1 and the carpal pad—that form the heel at the posterior edge of the track, and may or may not show. The negative space between toes and meta-carpals has little to no fur. Claws may or may not register in tracks. Front and hind tracks are similar dimensions.

Hind: 1¼–3½ in. long (3.2–8.9 cm) × ⅞–1¾ in. wide (2.2–4.4 cm)

Small to medium. Plantigrade. Slightly asymmetrical. Five toes in a classic rodent structure. Toes 1 and 5 point out to either side, and toes 2, 3, and 4 register close together, pointing forward. Some meta-tarsals are fused to form a larger, C-shaped pad, but four distinct pads are often evident in tracks. Additional pads are found in the heel portion of the track; during warmer summer months the heel sheds fur and is completely naked. Claws may or may not register and are much smaller than those of front feet.

TRAIL: Walk: Strides: 4–6 in. (10.2–15.2 cm)
Trail widths: 3¾–4¼ in. (9.5–10.8 cm)

Bound: Strides: 6–30 in. (15.2–76.2 cm) and occasionally more
Trail widths: 3¾–7 in. (9.5–17.8 cm)
Group lengths: 2–9 in. (5.1–22.9 cm)

NOTES:

- There are three large tree squirrels in California: the native Western Gray Squirrel (*S. griseus*) and two introduced species, the Eastern Gray (*S. carolinensis*) and Eastern Fox squirrels (*S. niger*). The Eastern Gray and Eastern Fox squirrels were largely introduced in urban centers along the coast and in the Central Valley. The fox squirrel is now numerous in urban and sub-urban settings, and both species are beginning to move into wilder country. For instance, in the Santa Cruz Mountains all three species are present.

- The tracks and trails of all three species are similar and easily confused. The exception is the tracks of adult Western Gray Squirrels, which are so large that they cannot be confused with the sign made by the slightly smaller Eastern Gray and Fox squirrels. As a suggested cutoff point, front tracks larger than 1¾ in. long (4.4 cm) × 1⅛ in. wide (2.9 cm) and hind tracks larger than 3¼ in. long (8.3 cm) × 1½ in. wide (3.8 cm) are almost certainly those of Western Grays.

- Tree squirrels bound when traveling and foraging on the ground. All three species also walk when foraging or scent-marking; however, Eastern Fox Squirrels walk much more frequently, and for longer stretches, than either Western or Eastern Gray squirrels.

- In deep fluffy snow the group lengths for bounds is reduced, and hind feet register next to or very close to the front tracks.

- In soft substrates the bound patterns may resemble those left by cottontails, however, their foot morphology is drastically different.

SCAT: ⅛–⁵⁄₁₆ (0.3–0.8 cm) diameter, ⁵⁄₃₂–³⁄₈ (0.4–1 cm) long.

The pellets of gray and fox squirrels, as opposed to red squirrels, seem to be dropped at random. We most often find them where squirrels have spent ample time feeding in a given area. Pellets are quite easy to find in new, wet snow. Pellets are dropped as squirrels travel overhead, and swell in the wet snow. You may find the bloated gray to brown pellets under tree canopies, sprinkled about. Those that soak up more snow may "explode," leaving a brown or gray stain with the residue of the pellet in the center.

URINE AND OTHER SCENT-MARKING BEHAVIORS: The scent marking of Western Gray Squirrels has yet to be studied with the intensity received by both Eastern Gray and Eastern Fox squirrels. Both Eastern Gray and Eastern Fox squirrels mark travel routes with quick nibbles, to scar the bark, and cheek wipes to deposit scent from the oral glands (Koprowski 1993). Males and females both exhibit this behavior, but males do so much more than females, and Eastern Fox Squirrels more frequently than Eastern Grays.

Both Eastern Gray and Eastern Fox squirrels also create large visual and olfactory scent posts at the bases of trees (often the lee side) and under large branches (Koprowski 1993). Elbroch (2003) refers to these sites as "territorial stripes." Unlike cheek wiping, marking at territorial stripes is strictly the realm of males. Marking was never observed for female fox squirrels in 89 instances, and only in two out of 85 observations were female gray squirrels seen marking at a territorial stripe (Koprowski 1993). At these marking sites squirrels slow down, pause, and intently smell the area before adding to the sign with cheek swipes and chewing the outer bark with the incisors. Elbroch (2003) presented data on 64 stripes chewed by Eastern Gray Squirrels on 37 species of trees, and found that 90 percent of the stripes fell between 13 and 70 in. (33–178 cm) of the ground.

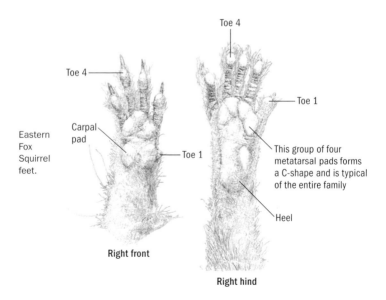

Toe 4

Toe 4

Toe 1

Eastern Fox Squirrel feet.

Carpal pad

Toe 1

This group of four metatarsal pads forms a C-shape and is typical of the entire family

Heel

Right front

Right hind

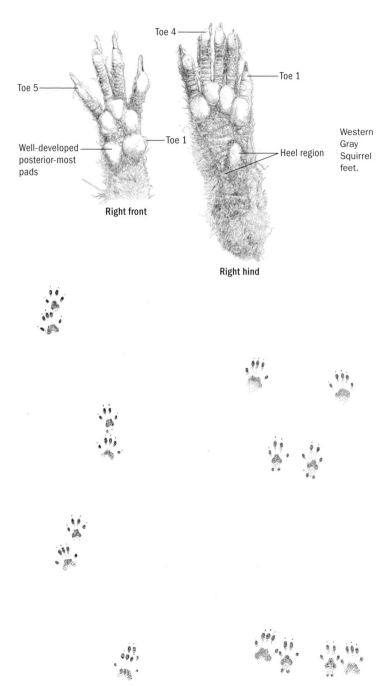

Toe 4

Toe 1

Toe 5

Well-developed posterior-most pads

Toe 1

Heel region

Western Gray Squirrel feet.

Right front

Right hind

Walking Eastern Fox Squirrel.

Bounding trail of large tree squirrels.

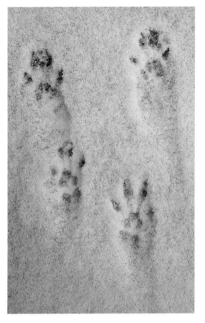

Eastern Gray Squirrel tracks in snow.

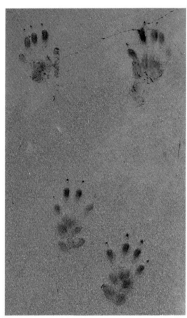

Western Gray Squirrel tracks in mud.

The walking trail of an Eastern Fox Squirrel.

An Eastern Fox Squirrel marking with its oral glands.

The territorial stripe of male Eastern Gray Squirrels.

DOUGLAS'S SQUIRREL   *Tamiasciurus douglasii*

TRACKS: Front: 1–1¼ in. long (2.5–3.2 cm) × ⁹/₁₆–1 in. wide (1.4–2.5 cm)

Small. Plantigrade. Asymmetrical. Five toes in classic rodent structure: toe 1 is a vestigial thumb; however, it is more developed than the toe 1 of larger tree squirrels and typically registers in tracks. Toes 2 and 5 point toward the sides, and toes 3 and 4 point forward. Some of the metacarpal pads are fused, but the front tracks include three distinct palm pads; there are two additional pads—the metacarpal pad for toe 1 and the carpal pad—that form the heel at the posterior edge of the track, and may or may not show. Unlike in larger tree squirrels, the digital pads often connect with the palm, giving them a fingerlike appearance. The negative space between toes and metacarpals has little to no fur. Claws may or may not register in tracks. Front and hind tracks are similar dimensions.

Hind: ⅞–2¼ in. long (2.2–5.7 cm) × ⅝–1⅛ in. wide (1.6–2.9 cm)

Small. Plantigrade. Slightly asymmetrical. Five toes in a classic rodent structure. Toes 1 and 5 point out to either side, and toes 2, 3, and 4 register close together, pointing forward. Some metatarsals are fused to form a larger, C-shaped pad, but four distinct pads are often evident in tracks. Although additional pads are found in the heel portion of the track, the heel has a fur covering and may or may not register. Claws may or may not register, and are much smaller than those of front feet.

TRAIL: Bound: Strides: 4–25 in. (10.2–63.5 cm) and occasionally more
Trail widths: 2⅞–4⅜ in. (7.3–11.1 cm)
Group lengths: 2–11 in. (5.1–27.9 cm)

Walk: Strides: 3½–5¼ in. (8.9–13.3 cm)
Trail widths: 2⅜–2¾ in. (6–7 cm)

NOTES:

• These medium-size arboreal squirrels are diurnal and bound for much of their time traveling on the ground.

• In deep fluffy snow the group length for bounds is reduced and hind feet register next to or very close to the front tracks. Douglas's Squirrels also form tunnels within the snow layer to travel in greater safety and to reach their stores of food at ground level.

SCAT: ³/₁₆–¼ in (0.5–0.6 cm) diameter, ³/₁₆–⁷/₁₆ in. (0.5–1.1 cm) long

The scats of Douglas's Squirrels may be dropped at random but are often placed carefully along travel routes, bridges over water and forest debris, and in other high-traffic areas. Like canines, high points in the "trail" are often marked, which is often a branch stub or burl on a limb.

Scats vary depending upon the moisture content in their diet. High moisture content results in soft scats, often twisted, and with tapered ends, or in clumps of "pellets" stuck together. More often we find small scats, reminiscent of deer pellets, in that they are often pointed at one end and have a dimple at the other. Moister scats will be less defined, more wrinkled, and usually pointy at both ends.

URINE AND OTHER SCENT-MARKING BEHAVIORS: Red Squirrels (*T. hudsonicus*), the eastern equivalent to the Douglas's Squirrel, bite and rub their oral glands along travel routes, especially on raised surfaces, such as burls or knobs, as well as at grooming and resting sites (Ferron and Ouellet 1989). Both males and females "cheek swipe," and researchers believe that the behavior may play the dual purpose of self-orientation for the animal (like landmarks) and advertisement of the occupancy of a territory.

Douglas's and Red squirrels also maintain regular scent sites, like grays, and these "territorial stripes" are more often horizontal than vertical along a dead lower branch of a dominant tree in the animal's territory (Elbroch 2003). These definitely play a territorial function. In addition, look for heavy biting and scent-marking close to occupied nests.

Bounding Douglas's Squirrel.   Douglas's Squirrel tracks in mud.

Douglas's Squirrel tracks in snow. Photo by Jonah Evans.

The horizontal biting and marking created by a Douglas's Squirrel.

NORTHERN FLYING SQUIRREL *Glaucomys sabrinus*

Plate 15
Plate 91

TRACKS: Front: $^{11}/_{16}$–1¼ in. long (1.7–3.2 cm) × ½–¾ in. wide (1.3–1.9 cm)

Small. Plantigrade. Asymmetrical. Five toes in classic rodent structure: toe 1 is a vestigial thumb and rarely registers in tracks. Toes 2 and 5 point toward the sides, and toes 3 and 4 point forward. Some of the metacarpal pads are fused, but three distinct palm pads appear in the front tracks; two additional pads, the metacarpal pad for toe 1 and the carpal pad, form the heel at the posterior edge of the track, and may or may not show. The negative space between the toes and metacarpal pads is heavily furred, obscuring the appearance of tracks. Claws may or may not register in tracks. The front tracks are smaller than the hind tracks.

Hind: 1¼–1⅞ in. long (3.2–4.8 cm) × ⅝–¹⁵⁄₁₆ in. wide (1.6–2.4 cm)

Small. Plantigrade. Slightly asymmetrical. Five toes in a classic rodent structure: toes 1 and 5 point out to either side, and toes 2, 3, and 4 register close together, pointing forward. However, toe 5 is similar in length to toes 2–4, and markedly longer than in other tree squirrels. Some metatarsals are fused to form a slender, C-shaped pad, somewhat obscured by the heavy fur on the foot. The heel is also covered in fur and may or may not register. Claws may or may not register, and are much smaller than those of front feet.

TRAIL: Bound: Strides: 6–34 in. (15.2–86.4 cm) and occasionally more
Trail widths: 2¾–4¼ in. (7–10.8 cm)
Group length: 2–7 in. (5.1–17.8 cm)

NOTES:

- Generally bound when traveling on the ground, but the front feet are often wider apart than in the trails of other tree squirrels, giving the groups of tracks a boxy appearance. Northern Flying Squirrels also occasionally hop.

- Very occasionally you will see marks made by the skin flaps in a bounding trail, but this is more often not the case.

- Trails may originate far from trees, where an animal had glided and landed. This landing spot is referred to by some authors as a "sitzmark" (Murie and Elbroch 2005). However, do not rely upon landing marks to identify flying squirrel trails, because it is more common for flying squirrels to climb down a tree to forage on the ground, than it is to glide down.

SCAT: ³⁄₃₂–³⁄₁₆ in. (0.2–0.5 cm) diameter, ⅛–⅜ in. (0.3–1 cm) long

Scats accumulate within hollow trees in which they are nesting, or beneath holes after they have cleaned within and evicted debris. Also investigate where predators such as martens or Fishers have excavated a hole in a standing snag, for they often pull out the nests and scats of mice and squirrels in the process.

URINE AND OTHER SCENT-MARKING BEHAVIORS: Like other sciurids, Northern Flying Squirrels use cheek swipes to scent-mark along regular travel routes (Ferron 1983). Ferron postulated that cheek swipes in this species may play a role in self-orientation while traveling about their territory, as well as contribute to time-share systems at grooming and resting sites used by more than one squirrel. Both males and females mark in this way.

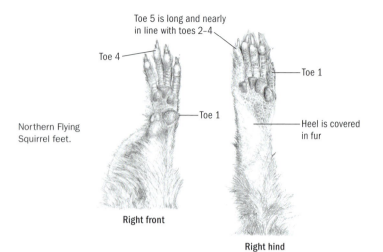

Toe 5 is long and nearly in line with toes 2–4

Toe 4

Toe 1

Toe 1

Heel is covered in fur

Northern Flying Squirrel feet.

Right front

Right hind

Northern Flying Squirrel bounding.

The landing mark, also called a "sitzmark," left by a gliding Northern Flying Squirrel.

Loping Northern
Flying Squirrel.

Bounding flying
squirrel.

WHITE-TAILED ANTELOPE SQUIRREL

Ammospermophilus leucurus

TRACKS: Front: $^5/_8$–$^{15}/_{16}$ in. long (1.6–2.4 cm) × $^3/_8$–$^5/_8$ in. wide (1–1.6 cm)

Tiny to small. Plantigrade. Asymmetrical. Five toes in classic rodent structure: toe 1 is a vestigial thumb and rarely registers in tracks. Toes 2 and 5 point toward the sides at 45-degree angles, and toes 3 and 4 point straight forward. Central metacarpal pads are fused, so look for three distinct palm pads; there are two additional pads, the metacarpal pad for toe 1 and the carpal pad, that form the heel at the posterior edge of the track, and tend to register clearly less frequently than the same pads in chipmunk tracks. The negative space between the toes and metacarpals has little to no fur. Claws are well developed (larger and longer than in the tracks of chipmunks and woodrats) and often prominent in tracks.

Hind: $^{11}/_{16}$–1 in. long (1.7–2.5 cm) × $^9/_{16}$–$^{13}/_{16}$ in. wide (1.4–2.1 cm)

Tiny to small. Plantigrade. Slightly asymmetrical. Five toes in a classic rodent structure. Toes 1 and 5 point out to either side, and toes 2, 3, and 4 register close together, pointing forward. Some metatarsals are fused to form larger pads, but four distinct pads will show in tracks. Claws may or may not register, and are much smaller than those of front feet. The heel portion of the foot may or may not register. Both the front and hind tracks are more asymmetrical than the similar-sized tracks of chipmunks or woodrats.

TRAIL: Bound: Strides: 5–15 in. (12.7–38.1 cm)
Trail widths: 2$^1/_8$–3½ in. (5.4–8.9 cm)
Group lengths: 2–4½ in. (5.1–11.4 cm)

NOTES:

• Front tracks are significantly smaller than hind ones, and this discrepancy is a key characteristic in identifying antelope squirrel sign.

• Move about in bounds and occasional hops; they walk when foraging or sunning themselves at burrow entrances.

• Associated with and adapted for dry habitats.

• They are competent climbers.

• Do not hibernate and remain active throughout the year.

SCAT: $^3/_{32}$–$^3/_{16}$ in. (0.2–0.5 cm) diameter

Scats are sprinkled in areas where they perch to vocalize and feed, while keeping an eye out for predators. Scats vary depending upon the moisture content in their diet. High moisture content results in soft scats, often twisted and with tapered ends, or in clumps of "pellets" stuck together. More often we find small scats reminiscent of deer pellets, in that they are often pointed at one end and have a dimple at the other. Moister scats will be less defined, more wrinkled, and usually pointy at both ends.

Bounding White-tailed
Antelope Squirrel.

The bounding tracks of a White-tailed
Antelope Squirrel in Death Valley National
Park.

YELLOW-BELLIED MARMOT   *Marmota flaviventris*

TRACKS: Front: 1¾–3 in. long (4.4–7.6 cm) × 1½–2⅜ in. wide (3.8–6 cm)

Small to medium. Plantigrade. Asymmetrical. Five toes in classic
rodent structure: toe 1 is a vestigial thumb that only registers in
the clearest tracks as a small protrusion on the inside of the inner,
posterior-most pad. Toes 2 and 5 point forward and outward, and

toes 3 and 4 point forward. The center metacarpal pads are fused but are still strongly lobed. Tracks are composed of three distinct palm pads, although it may at times appear to be four; two additional pads, the metacarpal pad for toe 1 and the carpal pad, form the heel at the posterior edge of the track, and may or may not show. All the pads of the feet are naked and have a bulbous appearance. Claws may or may not register in tracks. Front and hind tracks are of similar dimensions.

Hind: 1¾–3¼ in. long (4.4–8.3 cm) × 1⅝–2½ in. wide (4.1–6.4 cm)

Small to medium. Plantigrade. Slightly asymmetrical. Five toes in a classic rodent structure: toes 1 and 5 point out to either side, and toes 2, 3, and 4 register close together, pointing forward. Some metatarsals are fused to form larger pads, but four distinct pads show in tracks. There are two additional pads in the posterior of the track (the heel); the heel is hairless and may or may not register. Claws may or may not register, and are much smaller than those of front feet.

TRAIL: Walk: Strides: 3½–10 in. (8.9–25.4 cm)
              Trail widths: 4¼–7 in. (10.8–17.8 cm)

         Bound: Strides: 9–31 in. (22.9–78.7 cm)
              Trail widths: 4¾–7¾ in. (12.1–19.7 cm)
              Group lengths: 6–14 in. (15.2–35.6 cm)

NOTES:

• Walk for much of the time, switching to a bound when quickly retreating to a burrow system for safety or chasing an intruder from their territory. Trots and lopes are also used when crossing exposed areas, but when there is no imminent danger.

• Alpine animals, associated with rocky terrain, but will also be found in high elevation forests, living under boulders and in burrows

• Trails will begin and end in rock crevices or burrows. Runs are often apparent at burrow entrances.

• Hibernate for much of the year, slumbering through the harsh winter season. They reappear while snow still lingers, surviving on stores created the previous summer and autumn.

• Hind track can superficially resemble a raccoon track in some substrates, or more easily be confused with the track of a Western Gray Squirrel.

SCATS: ⅜–¹⁵/₁₆ in. (1–2.4 cm) diameter, 1½–3½ in. (3.8–8.9 cm) long, sometimes in chains much longer.

Marmot scats vary tremendously upon the moisture content of their diet. They vary from very soft, dark-green dollops to tubular scats of various colors. On occasion, great chains of linked pellets are produced, as well.

Scats are easy to find along travel routes, and near burrows and natural crevices they inhabit. Yellow-bellied Marmots create conspicuous and often large latrines atop boulders and other elevated surfaces.

Latrines may accumulate hundreds of defecations over time, and include bleached, aged scats from previous years.

URINE AND OTHER SCENT-MARKING BEHAVIORS: Marmots employ their oral glands in marking territories. They rub the sides of their mouths and cheeks on boulders, sticks, fence posts, tree trunks, human refuse, and shrubby branches of bushes, especially those near their burrow entrances. Marmots mark most often during spring and early summer, and then marking behaviors gradually decline through the summer. Male Yellow-bellied Marmots mark far more often than females, and can be seen investing good portions of their days in moving slowly about their range, marking and re-marking every available surface within 20 ft (6 m) of their burrows (Armitage 2003). In soft soils they also place the corner of their mouths in contact with the ground and then slide forward, propelled by their hind feet until they are flat out on the ground. They may push themselves forward a short distance, leaving troughlike trails like otters sliding in snow.

Walking Yellow-bellied Marmot.

Bounding Yellow-bellied Marmot.

The loping trail of a Yellow-bellied Marmot. Photo by Jonah Evans.

Left front (above) and hind tracks of a Yellow-bellied Marmot.

The accumulated scats at a large Yellow-bellied Marmot latrine in Tuolomne Meadows, Yosemite.

CALIFORNIA GROUND SQUIRREL   *Spermophilus beecheyi*

TRACKS: Front: ¹⁵/₁₆–1³/₈ in. long (2.4–3.5 cm) × ⁵/₈–⁷/₈ in. wide (1.6–2.2 cm)

Small. Plantigrade. Asymmetrical—more so than the tracks of tree squirrels. Five toes in classic rodent structure: toe 1 is a vestigial thumb. Toes 2 and 5 point forward and outward, and toes 3 and 4 point forward. Some of the metacarpal pads are fused, but front tracks reveal three distinct palm pads; two additional pads, the metacarpal pad for toe 1 and the carpal pad, form the heel at the posterior edge of the track, and may or may not show. The negative space between toes and metacarpals has little to no fur. Claws long and prominent in tracks. Front and hind tracks are of similar dimensions.

Hind: 1–1¹³/₁₆ in. long (2.5–4.6 cm) × ⁵/₈–1 in. wide (1.6–2.5 cm)

Small. Plantigrade. Slightly asymmetrical. Five toes in a classic rodent structure. Toes 1 and 5 point out to either side, and toes 2, 3, and 4 register close together, pointing forward. Although some metatarsals are fused to form larger pads, four distinct pads will show in tracks. Claws may or may not register, and are much smaller than those of front feet. The heel portion of the foot may or may not register, being naked and composed of several distinct pads.

TRAIL: Walk: Strides: 3¾–5¾ in. (9.5–14.6 cm)
                    Trail widths: 2¼–3½ in. (5.7–8.9 cm)

             Bound: Strides: 5–25 in. (12.7–63.5 cm)
                    Trail widths: 2⁷/₈–4½ in. (7.3–11.4 cm)
                    Group lengths: 2¾–8 in. (7–20.3 cm) ´

NOTES:

• Bound when traveling above ground. They walk when foraging and exploring.

• This large ground squirrel is diurnal and common throughout much of California. They inhabit open habitats, including agricultural lands, parks, and other human-altered habitats.

• Northern California Ground Squirrels hibernate for much of the year; in warmer southern habitats some hibernate little or not at all.

SCAT: ⁵/₃₂–⁵/₁₆ in. (0.4–0.8 cm) diameter, ¼–¾ in. (0.6–1.9 cm) long

Ground squirrel scats vary with the moisture content in their diet. High moisture content results in soft scats, often twisted, and with tapered ends; moist scats may also be clumped together. Scats also vary in shape and size. Drier scats are reminiscent of deer pellets, in that they are often pointed at one end and have a dimple at the other.

Scats of this species are easily found near burrow entrances. Also place scats in scrapes, like Golden-mantled Ground Squirrels.

URINE AND OTHER SCENT-MARKING BEHAVIORS: Chemical communication is widely used by ground squirrels, although the means vary across species. Many species establish and maintain territories by rubbing oral glands, located at the corners of their mouths, on the ground and available objects such as rocks and sticks (Kivett et al. 1976). California Ground Squirrels also scent-mark by dragging and rubbing their bodies over objects and along road edges—probably with anal glands.

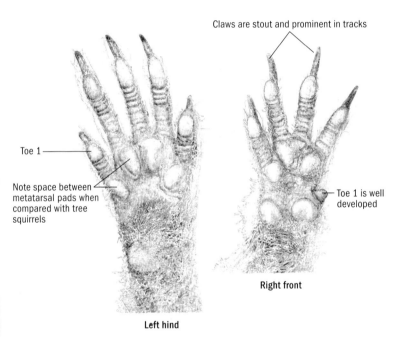

Claws are stout and prominent in tracks

Toe 1

Note space between metatarsal pads when compared with tree squirrels

Toe 1 is well developed

Right front

Left hind

California Ground Squirrel feet.

Bounding California Ground
Squirrel.

Trotting California
Ground Squirrel.

Note the well-pronounced asymmetry in the front tracks of this California Ground Squirrel.

The bounding trail of a California Ground Squirrel.

The trail of a California Ground Squirrel sliding through the dust to scent-mark.

BELDING'S GROUND SQUIRREL

Spermophilus beldingi

Plate 16

TRACKS: Front: $^{15}/_{16}$–1½ in. long (2.4–3.8 cm) × ½–$^{13}/_{16}$ in. wide (1.3–2.1 cm)

Small. Plantigrade. Asymmetrical—more so than the tracks of tree squirrels. Five toes in classic rodent structure: toe 1 is a vestigial thumb. Toes 2 and 5 point forward and outward, and toes 3 and 4 point forward. Some of the metacarpal pads are fused, but front tracks include three distinct palm pads; two additional pads, the metacarpal pad for toe 1 and the carpal pad, form the heel at the posterior edge of the track, and may or may not show. The negative space between toes and metacarpals has little to no fur. Claws long and prominent in tracks. Front and hind tracks are of similar dimensions.

Hind: $1^1/_{16}$–$1^9/_{16}$ in. long (2.7–4 cm) × $^{11}/_{16}$–$1^1/_{16}$ in. wide (1.7–2.7 cm)

Small. Plantigrade. Slightly asymmetrical. Five toes in a classic rodent structure: toes 1 and 5 point out to either side, and toes 2, 3, and 4 register close together, pointing forward. Some metatarsals are fused to form larger pads, but four distinct pads will show in tracks. Claws may or may not register and are much smaller than those of front feet. The heel portion of the foot, which is naked and composed of several distinct pads, may or may not register.

TRAIL: Bound: Strides: 4½–15 in. (11.4–38.1 cm)
Trail widths: $2^5/_8$–3½ in. (6.7–8.9 cm)
Group lengths: 3–5 in. (7.6–12.7 cm)

2 × 2 Lope/bound: Strides: 3–9 in. (7.6–22.9 cm)
Trail widths: 2½–4¼ in. (6.4–10.8 cm)

NOTES:

• Ground squirrels typically bound when traveling above ground. They will slow to a walk when foraging and occasionally use lopes to cross open areas at a medium pace.

• Ground squirrels are diurnal, and in California the Belding's Ground Squirrel is restricted to open habitats in alpine settings in the Sierra Nevadas and the high sagebrush habitats of the Great Basin.

• Belding's Ground Squirrels hibernate for much of the year.

SCAT: $^1/_8$–$^5/_{16}$ in. (0.3–0.8 cm) diameter, $^5/_{16}$–$^{13}/_{16}$ in. (0.8–2.1 cm) long

Ground squirrel scats vary with the moisture content in their diet. High moisture content results in soft scats, often twisted and with tapered ends; moist scats may also be clumped together. Scats also vary in shape and size. Drier scats are reminiscent of deer pellets, in that they are often pointed at one end and have a dimple at the other. Scats are easily found near burrow entrances.

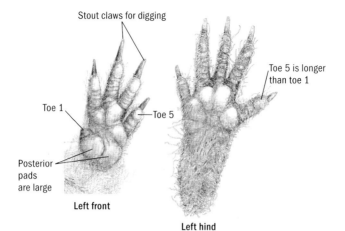

Stout claws for digging

Toe 1

Toe 5

Toe 5 is longer than toe 1

Posterior pads are large

Left front

Left hind

Belding's Ground Squirrel feet.

Belding's Ground Squirrel tracks in Tuolomne Meadows.

Loping Belding's Ground Squirrel.

GOLDEN-MANTLED GROUND SQUIRREL   *Spermophilus lateralis*

Plate 95

TRACKS: Front: ½–1 in. long (1.3–2.5 cm) × ½–⁹⁄₁₆ in. wide (1.3–1.4 cm)

Small. Plantigrade. Asymmetrical—more so than the tracks of tree squirrels. Five toes in classic rodent structure: toe 1 is a vestigial thumb. Toes 2 and 5 point toward the sides, and toes 3 and 4 point forward. Some of the metacarpal pads are fused, but front tracks include three distinct palm pads; there are two additional pads—the metacarpal pad for toe 1 and the carpal pad—that form the heel at the posterior edge of the track, and may or may not show. The negative space between toes and metacarpals has little to no fur. Claws long and prominent in tracks. Front and hind tracks are of similar dimensions. Tracks are more asymmetrical and toes are shorter than in chipmunks.

Hind: $^7/_8$–$1^3/_{16}$ in. long (2.2–3 cm) × $^9/_{16}$–$^{11}/_{16}$ in. wide (1.4–1.7 cm)

Small. Plantigrade. Slightly asymmetrical. Five toes in a classic rodent structure: toes 1 and 5 point out to either side, and toes 2, 3, and 4 register close together, pointing forward. Some metatarsals are fused to form larger pads, but four distinct pads will show in tracks. Claws may or may not register, and are much smaller than those of front feet. The naked heel portion of the foot is composed of several distinct pads that may or may not register.

TRAIL: Bound: Trail widths: 2¾ in. (7 cm)

NOTES:

- Bound when traveling above ground.
- Ground squirrels are diurnal, but unlike other ground squirrels, Golden-mantled species are associated with forested habitats. They also live very close to human settlements and are common in Sierra parks and campgrounds.

SCAT: $^3/_{16}$–$^5/_{16}$ in. (0.5–0.8 cm) diameter, $^3/_{16}$–$^1/_2$ in. (0.5–1.3 cm) long

Moisture content in the diet of ground squirrels is responsible for variation in their scats. High moisture content results in soft scats, often twisted and with tapered ends; moist scats may also be clumped together. Scats also vary in shape and size. Drier scats are reminiscent of deer pellets, in that they are often pointed at one end and have a dimple at the other.

Golden-mantled Ground Squirrels do not litter their burrow entrances and throw mounds with scats, as do other ground squirrels. Instead, look for scats deposited on prominent perches and along travel routes nearby.

URINE AND OTHER SCENT-MARKING BEHAVIORS: Golden-mantled Ground Squirrels create circular scrapes, between 1¼ and 2 in. (3.2–5.1 cm) in diameter, and urinate or defecate within them. Scent-marking probably plays a role in communicating occupancy of a territory (Ferron 1977).

Bounding tracks of a Golden-mantled Ground Squirrel in the high country of the Sierra Nevadas.

The scrape and scats of a Golden-mantled Ground Squirrel.

CHIPMUNKS *Tamias spp.*

TRACKS: Front: ³/₈–¹¹/₁₆ in. long (1–1.7 cm) × ⁵/₁₆–⁹/₁₆ in. wide (0.8–1.4 cm)

> Tiny. Plantigrade. Asymmetrical, but more symmetrical than similar-sized ground and antelope squirrels. Five toes in the classic rodent structure: toe 1 is a vestigial thumb, and rarely registers in tracks. Toes 2 and 5 angle toward the sides at 45 degrees, and toes 3 and 4 point straight forward. Central metacarpal pads are fused, so look for three distinct palm pads; two additional pads, the metacarpal pad for toe 1 and the carpal pad, form the heel at the posterior edge of the track, and may or may not show. The negative space between the toes and metacarpals has little to no fur. Claws may or may not register in tracks. Front and hind tracks are of similar dimensions.

> Hind: ½–¹¹/₁₆ in. long (1.3–1.7 cm) × ½–¹¹/₁₆ in. wide (1.3–1.7 cm)

> Tiny. Plantigrade. Slightly asymmetrical. Five toes in a classic rodent structure: toes 1 and 5 point out to either side at 45 degrees, and toes 2, 3, and 4 register close together and point straight forward. Although some metatarsals are fused to form larger pads, four distinct pads will show in tracks. Claws may or may not register, and are much smaller than those of the front feet. The heel portion of the foot is furred and may or may not register.

TRAIL: Bound: Strides: 3–20 in. (7.6–50.8 cm)
Trail widths: 1½–2¾ in. (3.8–7 cm)
Group lengths: 1¾–5 in. (4.4–12.7 cm)

NOTES:

- There are 13 species of chipmunks in California, some of which use distinctive habitats in specific regions of the state; these can be identified to species by using known information on their ranges. Others overlap in range with other chipmunks, making identification to the species level very difficult, if not impossible. The tracks and signs for all of the species are nearly identical, excepting that some species are slightly larger than others.

- Chipmunks move about in bounds, with occasional hops when they are moving very slowly. Short walks may be used when foraging or tracking down an odor by scent.

- Although chipmunks are not true hibernators, chipmunks stay underground for much of the snow season. However, it is not unusual to find them out and about during warm spells or as daylight hours increase noticeably near the end of winter.

- Chipmunks are competent climbers and will be found harvesting fruits and nuts above ground in trees and shrubs.

- Trails will begin and end with a burrow, rock crevice, or disappearing into thick cover, such as a slash pile.

- In deep fluffy snow, chipmunks will use a modified 2 × 2 bound, in which the front and hind tracks are found together in the same holes.

SCAT: $^3/_{32}$–$^5/_{32}$ in. (0.2–0.4 cm) diameter, $^1/_8$–$^9/_{32}$ in. (0.3–0.7 cm) long

Like other rodents, chipmunk scats vary depending upon the moisture in and content of their diet, but are often more spherical than other sciurids. They can be soft and pointy and stick together in clumps, or create individual pellets like those of squirrels, and on occasion, be linked. Chipmunk scats are often found at regular perches in their home ranges. Scent-marking in chipmunks has been little studied.

URINE AND OTHER SCENT-MARKING BEHAVIORS: Least Chipmunks (*T. minimus*) sometimes deposit urine on buried patches of seeds that are either exhausted or of poor quality to improve their foraging efficiency (Devenport et al. 1999).

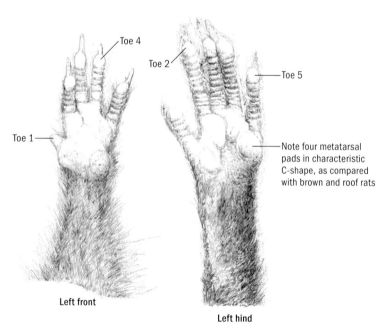

Merriam's Chipmunk feet.

Bounding Merriam's Chipmunk.

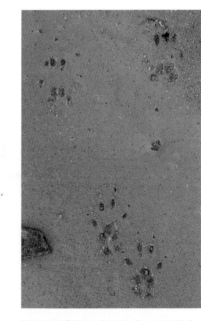

Merriam's Chipmunk tracks in a stretched-out bound.

Merriam's Chipmunk tracks, in a slower bound, and accompanying mice tracks.

Family Castoridae

AMERICAN BEAVER   *Castor canadensis*

Plate 32

Plates 100 and 115

TRACKS: Front: 2½–3⅞ in. long (6.4–9.8 cm) × 2¼–3½ in. wide (5.7–8.9 cm)

Medium. Plantigrade. Asymmetrical. Five toes in classic rodent structure: toe 1 is reduced, clawed, and occasionally registers clearly in tracks. Toes 2 and 5 point toward the sides, and toes 3 and 4 point forward. All toes are strongly curved toward the inside of the trail. Some of the metacarpal pads are fused; there are two additional pads—the metacarpal pad for toe 1 and the carpal pad—that form the heel at the posterior edge of the track, and may or may not show. Claws are long, blunt, and prominent in tracks. The front tracks are significantly smaller than the hind ones, and are often lost in the trail, crushed by the hind feet or tail that follow.

Hind: 4¾–7 in. long (12.1–17.8 cm) × 3¼–5¼ in. wide (8.3–13.3 cm)

Large to very large. Plantigrade. Asymmetrical. Five long toes. Toes 1 and 2 are not always obvious in tracks, while toes 3–5 register reliably. Distal webbing may or may not register. Metatarsal pads are fused to form larger pads, but these may or may not be distinct, depending upon substrate. Claws are large, short, and blunt. The rounded heel almost always registers and is an important feature used in identification. Nutria tracks are more slender, have sharp claws, and tend not to register the heel portion of the hind feet.

TRAIL: Walk: Strides: 6–11½ in. (15.2–29.2 cm)
Trail widths: 5¾–11 in. (14.6–27.9 cm)

Bound: Strides: 10–32 in. (25.4–81.3 cm)
Trail widths: 6¾–13½ in. (17.1–34.3 cm)
Group lengths: 7–14½ in. (17.8–36.8 cm)

NOTES:

• Walk when traveling and foraging. When threatened or alarmed, they will bound to the safety of water.

• Although much of their time is spent in the water, beaver trails are commonly found leaving water sources to forage, scent-mark, and occasionally travel to separate water sources.

• Do not climb but will occasionally shuffle up leaning logs and branches.

SCAT: ¾–1½ in. (1.9–3.8 cm) diameter, 1¼–3 in. (3.2–7.6 cm) long

Scats are often composed of three to five separate "pellets," or may be tubular, and are always filled with tiny bits of woody materials and other plant fibers. Beavers typically defecate in the water, but occasionally on land as well. Beavers also form latrines in secluded spots such as inlets and small, shallow ponds, some of which may dry up seasonally to reveal scats. One latrine in a shallow stream inlet held approximately 30 scats (Elbroch 2003).

Like rabbits, beavers engage in coprophagy and ingest scats to better break down the cellulose in vegetation. Their initial scats are soft and green, but we are unlikely to find them without disturbing a beaver during the process (Wilsson 1971).

URINE AND OTHER SCENT-MARKING BEHAVIORS: Beavers also communicate using scent from two sources, the anal scent glands and the castor sacs, both of which are located just inside the anus. Secretions from anal scent glands may be deposited alone or in combination with castor secretions, and have been shown to communicate information on kinship and gender (Müller-Schwarze and Heckman 1980). Castoreum is created when urine mixes with organic compounds in the castor sacs, and is deposited on grass tufts, mounds created from debris dragged up from the bottom of their ponds, or occasionally, dry materials scraped together on land. Once a mound is created, and some are over a foot tall, a beaver walks over it and drags its anus across the top to deposit castoreum (Svendsen 1978). (See the photo of a scent mound on page 104.)

Scent mounds are primarily thought to mark territorial boundaries, because beavers can identify the individual that constructed them (Svendsen 1980). Scent mounds are also most often constructed and maintained in late winter and spring, when transient beavers are roaming about looking to establish their own territories. A beaver may create more than 100 mounds within its territory along dams, trails, and the shores of its ponds and rivers. Males create scent mounds far more frequently than females, although all members of the colony create them.

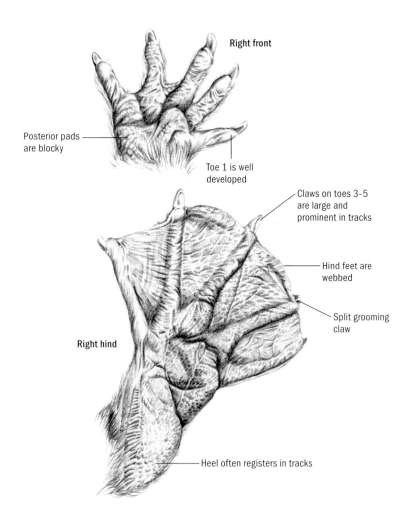

Right front

Posterior pads are blocky

Toe 1 is well developed

Claws on toes 3–5 are large and prominent in tracks

Hind feet are webbed

Split grooming claw

Right hind

Heel often registers in tracks

American Beaver feet.

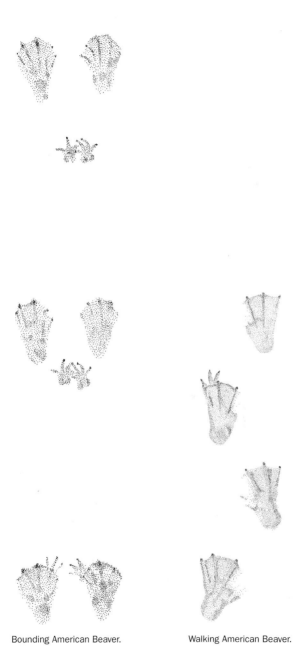

Bounding American Beaver.

Walking American Beaver.

American Beaver tracks in Davis, California. The much smaller left front track sits below and the larger left hind track above. In the lower right corner of the photograph is a raccoon track.

The walking trail of an American Beaver.

Pocket Mice and Kangaroo Rats: Family Heteromyidae

POCKET MICE   *Perognathus* and *Chaetodipus* **spp.**

Plate 13

TRACKS: Front: ¼–³/₈ in. long (0.6–1 cm) × ¼–³/₈ in. wide (0.6–1 cm)

Tiny. Plantigrade. Asymmetrical. Five toes in classic rodent structure: toe 1 is a vestigial thumb and rarely registers in tracks. Toes 2 and 5 point toward the sides, and toes 3 and 4 point straight forward. The metacarpal region is covered in fur, but pads are often evident. Claws are long and generally clear in tracks. Front tracks are significantly smaller than hind tracks, a distinguishing feature for this family of animals.

Hind: ⁵/₁₆–½ in. long (0.8–1.3 cm) × ¼–³/₈ in. wide (0.6–1 cm)

Tiny. Plantigrade. Asymmetrical. Five toes. Toe 1 is greatly reduced and easily overlooked in the track; toes 2, 3, and 4 are long and often curved in toward the center of the trail. When toes 2–4 are the only obvious sign, they might be confused with tiny bird tracks. There is often a large negative space between the toes and the metatarsal region. The metatarsal region is covered in fur and often indistinct. Claws often apparent. The heel may or may not register.

TRAIL: Bound: Strides: 2¼–6 in. (5.7–15.2 cm)
Trail widths: ¾–1¼ in. (1.9–3.2 cm)
Group lengths: 1–3¼ in. (2.5–8.3 cm)

NOTES:

• The nine species of pocket mice in California leave very similar sign. In some areas of the state, specific species are habitat specialists, allowing you to narrow the potential perpetrator to species or to just several known species. In addition, there is considerable variation in size between the species. For instance, in the Los Padres National Forest in southern California, the trail of the larger California Pocket Mouse (*Chaetodipus californicus*) stands out from the smaller *Perognathus* species with which it overlaps in range.

• Pocket mice bound for travel purposes when foraging and exploring, and a key trail characteristic is that the front feet often land adjacent and overlap or are in contact with each other.

• Pocket mice remain dormant within burrows for as much as nine months of the year, surviving on seed caches between long naps at reduced body temperatures.

• Pocket mice are closely related to kangaroo rats, and in some substrates their hind tracks look like miniature versions of kangaroo rat hind tracks.

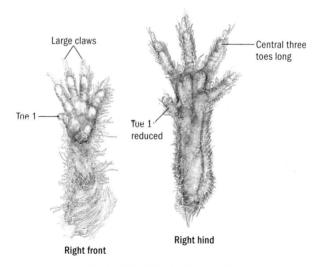

Large claws

Central three toes long

Toe 1

Toe 1 reduced

Right hind

Right front

Baja (or Bailey's) Pocket Mouse feet.

Pocket Mice and Kangaroo Rats: Family Heteromyidae

Pristine pocket mice tracks in mud. Note the huge difference in size between the inner front tracks and the outer hind tracks. Photo by Jonah Evans.

Pocket mice tracks. Look closely and you'll be able to make out the innermost toes in the hind tracks.

Bounding pocket mouse.

Small and Medium Kangaroo Rats   (*Dipodomys* spp.)

Plate 13

Plates 89 and 90

TRACKS: Front: ³/₈–⁹/₁₆ in. long (1–1.4 cm) × ³/₁₆–³/₈ in. wide (0.5–1 cm)

Tiny. Plantigrade. Asymmetrical. Five toes. Toe 1 is a vestigial thumb. Toes 2 and 5 point toward the sides, and toes 3 and 4 point forward. Central metacarpal pads are fused; there are two additional pads (the heel) at the posterior edge of the track, which may or may not show. Claws are long and an important feature of tracks. Front tracks are significantly smaller than hind ones. Front feet only occasionally touch the ground and leave tracks.

Hind: ¹¹/₁₆–1½ in. long (1.7–3.8 cm) × ³/₈–¹⁵/₁₆ in. wide (1–2.4 cm)

Tiny to small. Plantigrade. Asymmetrical. Four or five toes. Toe 1, when present, is a vestigial thumb, and toes 2–5 are very long and register close together. Metatarsal region furred and often indistinct. Claws are often apparent but are proportionately smaller than those of front feet. The long heel may or may not register.

TRAIL: Bipedal hop: Strides: 2–38 in. (5.1–96.5 cm)
Trail widths: ⁹/₁₆–2 in. (1.4–5.1 cm)

NOTES:

- We describe 12 of the 14 species of kangaroo rats in California as small or medium in size (see next account). Habitat and distribution within the state are useful in narrowing to species identification; however, where two or more species occur together, you may only be able to identify the sign to genus.

- Note that mammalogists use the number of toes on the hind feet to help distinguish among kangaroo rats, because some species have five toes and others have four. In tracking this is unfortunately less useful, because the tiny fifth toe, when present on the foot, rarely registers in the actual footprint—and you might need a magnifying glass to see it if it did.

- Kangaroo rats use a bipedal hop when traveling and exploring on the landscape. When threatened or alarmed, their strides increase significantly, and/or they may employ a skip. When foraging or investigating an area intensely, kangaroo rats use a slow bipedal hop or bound on all four feet. They drag their tail on the ground when moving on all four feet, but in faster gaits it is held aloft and rarely leaves a mark.

- Kangaroo rat burrows are conspicuous in areas they inhabit.

SCAT: ¹/₁₆–¹/₈ in. (0.2–0.3 cm) diameter, ¼–½ in. (0.6–1.3 cm) long

Scats are often longer than wide, smooth, and slightly curved like a kidney bean. Scats can be found near burrow entrances and dropped at random in areas of heavy foraging.

URINE AND OTHER SCENT-MARKING BEHAVIORS: Kangaroo rats have acute senses of smell and employ scent-marking as a nonconfrontational means of communicating with other kangaroo rats. One method is sand or dust bathing. When sand bathing, they till up the earth's surface by digging with their forepaws and then slide into the sand, like otters into water. They start with a cheek and run the length of one side of their bodies across the ground. They also drag their anuses along the ground near their burrows or before engaging in a sand bath to deposit secretions from their perineal glands. Kangaroo rats also urinate regularly at scent posts, and Desert Kangaroo Rats create latrines where scats accumulate. At latrines they often rub their cheeks and sides on nearby objects, whether they be the dead branches of a shrub or some novel item such as an abandoned glove. All of these marked areas provide sites where information on neighboring kangaroo rats, including sexual status and availability, can be checked without direct communication or conflict between them (Eisenberg 1963).

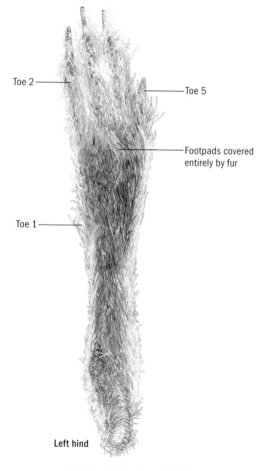

Toe 2

Toe 5

Footpads covered entirely by fur

Toe 1

Left hind

Heermann's Kangaroo Rat foot.

Another variation of bipedal
hopping in kangaroo rats.

Bipedal hopping
kangaroo rat in dust.

Four paired sets of hind tracks of Heerman's Kangaroo Rats, three moving up the photo and the most subtle moving down in the center of the photograph. Look carefully for the impressions made by the fur on their soles and by their claws.

In deep splayed tracks, the innermost toes registered clearly in these hind tracks of an Ord's Kangaroo Rat, *D. ordii*.

Large Kangaroo Rats

DESERT KANGAROO RAT   *Dipodomys deserti*

TRACKS: Front: ⅝–¾ in. long (1.6–1.9 cm) × ⅜–½ in. wide (1–1.3 cm)

Tiny. Plantigrade. Asymmetrical. Five toes. Toe one is a vestigial thumb. Toes 2 and 5 point toward the sides, and toes 3 and 4 point forward. Central metacarpal pads are fused; there are two additional pads (the heel), at the posterior edge of the track, which may or may not show. Claws are long and an important feature of tracks. Front tracks are significantly smaller than hind tracks. Front feet only occasionally touch the ground and leave tracks.

Hind: 1⅛–2³⁄₁₆ in. long (2.9–5.6 cm) × ¹¹⁄₁₆–1⅛ in. wide (1.7–2.9 cm)

Small. Plantigrade. Asymmetrical. Four toes. Toe 1 is absent, and the remaining, long toes register close together. Metatarsal region is furred and often indistinct. Claws are often apparent but are smaller than those of front feet. The long heel portion of the foot, which is furred, may or may not register.

TRAIL: Slow bipedal hop: Strides: 2–10 in. (5.1–25.4 cm)
                          Trail widths: 2½–3 in. (6.4–7.6 cm)

Bipedal hop/skip: Strides: 6–54 in. (15.2–137.2 cm), occasionally longer
                          Trail widths: 1⅜–2⅝ in. (3.5–6.7 cm)

NOTES:

- We describe two species of kangaroo rats in California as large: the endangered Giant Kangaroo Rat (*Dipodomys ingens*) only inhabits a very small area in southern California, while the more common Desert Kangaroo Rat (*D. deserti*) lives in the sand dunes in the southwest of the state, and other surrounding desert habitats.

- Kangaroo rats use a bipedal hop when traveling and exploring on the landscape. When threatened or alarmed, their strides increase significantly, and/or they may use a skip. When foraging or investigating an area intensely, kangaroo rats use a slow bipedal hop or bound on all four feet. They drag their tails on the ground when moving on all four feet, but in faster gaits it is held aloft and rarely leaves a mark.

- Kangaroo rat burrows are conspicuous in areas they inhabit.

SCAT: Scat: ¹⁄₁₆–³⁄₁₆ in. (0.2–0.4 cm) diameter, ¼–½ in. (0.6–1.3 cm) long

Scats are often longer than wide, smooth, and slightly curved like a kidney bean. Scats can be found near burrow entrances and dropped at random in areas of heavy foraging. Of California's kangaroo rats, the Desert Kangaroo Rat forms some of the more conspicuous latrines, often where sticks or other objects provide a centerpiece for marking with their oral glands as well.

URINE AND OTHER SCENT-MARKING BEHAVIORS: These behaviors are discussed in depth in the previous account (small and medium kangaroo rats). Mark with oral glands, anal drags, and through dust bathing.

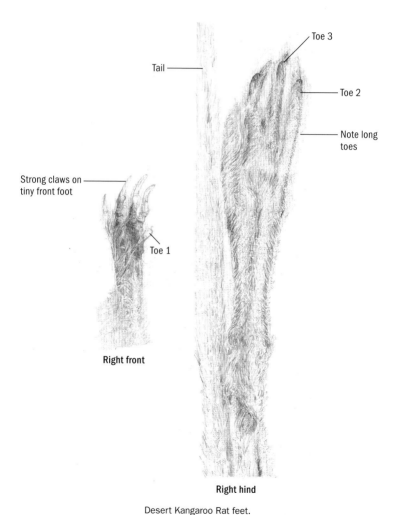

Desert Kangaroo Rat feet.

Bipedal hopping
of a Desert
Kangaroo Rat.

Skipping Desert
Kangaroo Rat.

Bipedal trails of hopping Desert Kangaroo Rats.

The latrine of a Desert Kangaroo Rat adjacent to dried twigs that the animal scent-marks with oral glands.

The long, slithering trail of a Desert Kangaroo Rat sliding across the sand as it scent-marks after emerging from its burrow. Most such sign is shorter than this.

Pocket Gophers: Family Geomyidae

POCKET GOPHERS   *Thomomys* **spp.**

Plate 14
Plate 92

TRACKS: Front: ⁷/₈–1⁵/₁₆ in. long (2.2–3.3 cm) × ³/₈–¹¹/₁₆ in. wide (1–1.7 cm)

Small. Plantigrade. Asymmetrical. Five toes. Toe 1 is reduced, but is clawed and still registers. Toes 1 and 5 are shorter than toes 2–4, and are located more to the posterior of the track. Toes 2, 3, and 4 point forward. Some of the metacarpal pads are fused, but tracks include three distinct palm pads; an additional metacarpal pad and the carpal pad form the heel at the posterior edge of the track. The long claws are a significant feature of tracks. Front track is narrower and longer than hind track.

Hind: $^5/_8$–1¼ in. long (1.6–3.2 cm) × $^7/_{16}$–$^{11}/_{16}$ in. wide (1.1–1.7 cm)

Small. Plantigrade. Slightly asymmetrical. Five toes in a classic rodent structure: toes 1 and 5 are short, register behind toes 2–4, and point out to either side. Toes 2, 3, and 4 register close together, pointing forward. Some metatarsals are fused, but distinct pads are often apparent. Claws register well in tracks. The heel may or may not register.

TRAIL:
Walk: Strides: 1$^3/_8$–3 in. (3.5–7.6 cm)
           Trail widths: 1$^1/_8$–2 in. (2.9–5.1 cm)

Trot: Strides: 3½–4$^1/_8$ in. (8.9–10.5 cm)
           Trail widths: 1$^3/_8$–1½ in. (3.5–3.8 cm)

NOTES:

• Walk about above the ground and trot when feeling exposed or traveling long distances (as during dispersal events). On occasion, they also bound.

• In deep snow, pocket gophers also move along the ground surface within the snow layer—especially evident in spring when dirt tubes cover landscapes after snows have receded.

SCAT: $^1/_8$–$^3/_{16}$ in. (0.3–0.5 cm) diameter, $^5/_{16}$–$^7/_{16}$ in. (0.8–1.1 cm) long

Scats are capsule-shaped or slightly curved, with smooth surfaces and rounded ends. Contents of scats we've seen are woody, but summer scats are probably softer and vary more in shape.

Pocket gopher scat is not common above ground but can be found if you look for it, especially just after snows recede. During the winter, Botta's Pocket Gophers create latrines above ground near a nest—often tunnel-shaped because they backfill an excavation in the snow with their feces. In addition, you might find mounds that seem to be the result of some sort of spring cleaning, where scats and other odds and ends have been pushed out by the gopher and the entrance plugged again.

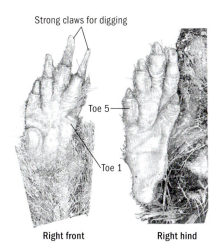

Strong claws for digging

Toe 5

Toe 1

Right front

Right hind

Botta's Pocket Gopher feet.

Walking pocket gopher.

The overstep straddle trot of a Botta's Pocket Gopher in the Los Padres National Forest, California.

The walking trail of a Northern Pocket Gopher.

The latrine of a Botta's Pocket Gopher in Cuyamaca Rancho State Park in San Diego County, exposed after winter snows receded.

Jumping Mice: Family Dipodidae

JUMPING MICE *Zapus* spp.

TRACKS: Front: $^3/_8$–$^5/_8$ in. long (1–1.6 cm) × $^3/_8$–$^5/_8$ in. wide (1–1.6 cm)

Tiny. Plantigrade. Asymmetrical. Five toes in classic rodent structure: toe 1 is a vestigial thumb and rarely registers in tracks. Four toes are long and slender: toes 2 and 5 point toward the sides or curve backward, and toes 3 and 4 point forward. Some of the metacarpal pads are fused, but tracks include three distinct palm pads; an additional metacarpal pad and the carpal pad form the heel at the posterior edge of the track. Claws may or may not register. Front tracks are significantly smaller than hind ones and are angled outward.

Hind: $^7/_{16}$–$^7/_8$ in. long (1.1–2.2 cm) × $^3/_8$–$^{11}/_{16}$ in. wide (1–1.7 cm)

Tiny. Plantigrade. Slightly asymmetrical. Five slender, long toes in a classic rodent structure: toe 1, which is very small, and toe 5 point out to either side, and are significantly shorter than toes 2, 3, and 4, which point forward. Toes 2 and 4 sometimes bow outward as well. Toes 2, 3, and 4 connect to the top of the metatarsal region, while toes 1 and 5 connect to the lower sides; this creates a slender "neck" in the track just posterior to the central three toes and in front of toes 1 and 5. Toes long, connected to the palm, and ribbed. Some metatarsals are fused to form larger pads, but distinct pads sometimes show in tracks. Claws more prominent than in other mice tracks. Contrary to previous literature (Murie 1954), the heel portions of feet rarely register in jumping mice tracks.

TRAIL: Bound: Strides: 4–48 in. (10.2–121.9 cm) (Jumping mice can bound up to 12 ft [365.8 cm].)
Trail widths: 1$^3/_8$–2$^1/_4$ in. (3.5–5.7 cm)
Group lengths: 1–2$^1/_2$ in. (2.5–6.4 cm)

NOTES:

- Bound when traveling and moving about. On very rare occasions they will leave a walking trail.
- Hibernators, and their tracks will not be seen for approximately six months of the year in the northern parts of their range.
- Trails are especially common in riparian areas. They are capable swimmers, and sometimes take to water as a means of predator evasion.

SCAT: $^1/_{16}$–$^1/_8$ in. (0.2–0.3 cm) diameter, $^1/_8$–$^1/_2$ in. (0.3–1.3 cm) long.

Scats are twisted and with tapered ends. We have never identified jumping mouse scat in the wild.

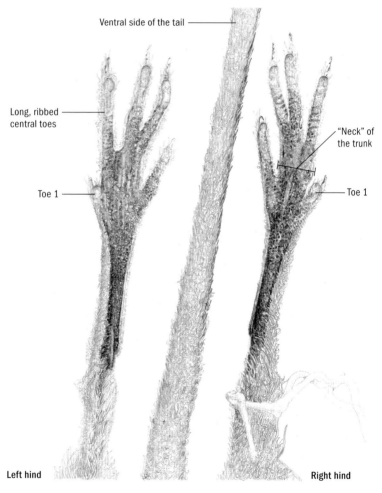

Ventral side of the tail

Long, ribbed
central toes

"Neck" of
the trunk

Toe 1

Toe 1

Left hind

Right hind

Western Jumping Mouse feet.

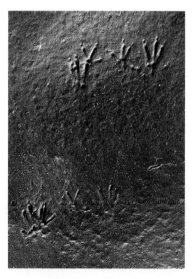

Bounding jumping mouse.

Tracks of a Western Jumping Mouse.

Native Rats, Mice, and Voles: Family Cricetidae

VOLES *Microtus* **and** *Myodes* **spp.**

Plate 1
Plate 13
Plate 89

TRACKS: Front: ¼–½ in. long (0.6–1.3 cm) × ⁹/₃₂–⁹/₁₆ in. wide (0.7–1.4 cm)

Tiny. Plantigrade. Asymmetrical. Five toes in classic rodent structure: toe 1 is greatly reduced. Toes 2 and 5 angle toward the sides (sometimes forming a straight line), and toes 3 and 4 point forward. The digital pads are less bulbous than those of mice. The slender toes are ribbed and often connected to the palm. Some of the metacarpal pads are fused, but there are three distinct palm pads; an additional metacarpal pad and the carpal pad form the heel at the posterior edge of the track, and they often register more weakly than those of White-Footed Mice, if at all. Claws may or may not register. The front tracks are smaller than the hind ones.

Hind: ¼–¹¹/₁₆ in. long (0.6–1.7 cm) × ¼–⁵/₈ in. wide (0.6–1.6 cm)

Tiny. Plantigrade. Slightly asymmetrical. Five long toes in a classic rodent structure: toes 1 and 5 angle to the sides (sometimes forming a straight line), and toes 2, 3, and 4 point forward. Slender, ribbed toes often register as connected to the palm. There is partial fusing of some metatarsal pads, but four palm pads should be distinct as well as two additional heel pads at the posterior of the track: the impression of the central palm pads often forms an upside down "L" and "·", the dot forming the corner of the rectangle if the "L" and "·" were combined. Note that the rectangle created by the "L"–"·" combination is longer than wide, whereas in deer mice it is wider than long. Claws and heel may or may not register.

TRAIL:

Trot: Strides: 2–3 ¼ in. (5.1–8.3 cm)
       Trail widths: ⁷/₈–1¹¹/₁₆ in. (2.2–4.3 cm)

Walk: Strides: 1¼–1¹⁵/₁₆ in. (3.2–4.9 cm)
       Trail widths: 1¼–2 in. (3.2–5.1 cm)

Bound: Strides: 4–9¼ in. (10.2–23.5 cm)
       Trail widths: 1³/₁₆–1½ in. (3–3.8 cm)

2 × 2: Strides 3–7 in. (7.6–17.8 cm)
       Trail widths: 1–1³/₈ in. (2.5–3.5 cm)

Hop: Strides: 5½–7 in. (14–17.8 cm)
       Trail widths: 1³/₈–1½ in. (3.5–3.8 cm)
       Group lengths: 1½–2 in. (3.8–5.1 cm)

NOTES:

• Medium to large vole species almost always travel in a direct-register trot, speeding up into hops, lopes, and bounds when exposed or feeling threatened. They also forage in an understep walk.

• Voles tend to remain within the snow layer when depths accumulate.

SCAT: $^1/_{16}$–$^1/_8$ in. (0.2–0.3 cm) diameter, $^1/_8$–$^5/_{16}$ in. (0.4–0.8 cm) long

Vole scat is capsule-shaped, with smooth surfaces and rounded ends; they are very symmetrical and regular. Vole scats can be found anywhere they have spent time, but they accumulate in latrines at dead-end runs, near nests, and at trail junctions along runs. Novel objects are often marked with feces, and researchers exploit this behavior to estimate vole numbers. They place small tile squares at intervals along trails and return up to a week later to count the scats placed upon them.

URINE AND OTHER SCENT-MARKING BEHAVIORS: Scent communication is highly developed in many voles and other species. Sexual status, health, and age and diet information are all included in scats, urine, and secretions left when rubbing the anogenital region (Ferkin et al. 2001). When a meadow vole, a close relative of the California Vole, discovers a scent mark of the other sex, it will often mark atop or adjacent to it with scents of its own. In this way males and females keep tabs on each other without actually meeting, and when a female is in heat, males use her scent marks to track her down.

Glandular secretions are difficult to see in the field, and so is urine in anything but the driest conditions. Calcareous (calcium-rich) deposits formed by accumulating vole urine can be found in habitats and seasons of infrequent rains, especially common for California Voles. They might be found at cracks in the earth where they move underground, burrow entrances, as well as upon raised objects or earth, where multiple animals mark repeatedly over time.

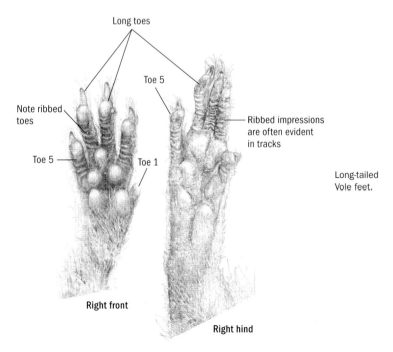

Long toes

Toe 5

Note ribbed toes

Ribbed impressions are often evident in tracks

Toe 5

Toe 1

Long-tailed Vole feet.

Right front

Right hind

Trotting vole.   Hopping vole.

The trail of a trotting California Vole, crisscrossed by Spotted Sandpiper and Western Gray Squirrel trails.

A California Vole latrine and accompanying calcareous deposits near Monterey, formed through repeated scent-marking in the dry season.

Clear front and hind tracks of a vole.

MUSKRAT

Ondatra zibethicus

Plate 21
Plates 97 and 118

TRACKS: Front: ⁷/₈–1½ in. long (2.2–3.8 cm) × 1–1½ in. wide (2.5–3.8 cm)

Small. Plantigrade. Asymmetrical. Five toes in classic rodent structure: toe 1 is greatly reduced but has a claw. Toes 2 and 5 angle toward the sides, and toes 3 and 4 point forward. The digital pads are somewhat bulbous, but the toes are slender and often connect to the palm. Some of the metacarpal pads are fused, but there are three distinct palm pads; an additional metacarpal pad and the carpal pad form the heel at the posterior edge of the track, and they are often visible in the track. The claws are long and prominent in tracks. The front tracks are significantly smaller than the hind tracks.

Hind: 1½–2¾ in. long (3.8–7 cm) × 1½–2½ in. wide (3.8–6.4 cm)

Small to medium. Plantigrade. Asymmetrical and distinctive. Five long toes in a rodent structure: toes 1 and 5 angle to the sides, and toes 2, 3, and 4 register closer together and are of similar dimensions. The toes are wide, ribbed, and connected to the palm; they are also surrounded by a "shelf" created from stiff hairs that aid the animal in swimming. The entire outline of the foot tends to register in tracks, obscuring individual metatarsal pads, although some can be differentiated in good substrate. There are two additional heel pads at the posterior of the track, which rarely register. The claws tend to register. Hind tracks tend to register at a severe angle, pointing inward.

TRAIL: Direct-register walk/overstep: Strides: 3–7 in. (7.6–17.8 cm)
Trail widths: 3–5 in. (7.6–12.7 cm)

Lope: Strides: 1⅝–4½ in. (4.1–11.4 cm)
Group lengths: 7¼–9¾ in. (18.4–24.8 cm)

Hop: Strides: 6–17 in. (15.2–43.2 cm)
Group lengths: 3½–6½ in. (8.9–16.5 cm)

NOTES:

- Walk when traveling and exploring, but speed up into hops and lopes when feeling exposed or threatened.

- Associated with riparian areas of every kind, including saltwater marshes, tidal inlets, and even coastal islands.

- In some areas, Muskrats use burrow systems rather than build lodges, and regular runs can be seen connecting these to foraging areas.

SCAT: Hard: ³⁄₁₆–⁵⁄₁₆ in. (0.5–0.8 cm) diameter, ⅜–1 in. (1–2.5 cm) long

Soft: ⅜–1 in. (1–2.5 cm) diameter, ⅞–1¾ in. (2.2–4.4 cm) long

Scats vary with the moisture content in their diet. May be soft and amorphous, clumped "pellets" stuck together, or harder, individual "pellets." Muskrats form scent posts, where tremendous amounts of scat may accumulate. Look for them along their runs, as well as on prominent elevated surfaces in and around water sources, such as rocks, grass clumps, or partially submerged logs. On occasion, Muskrats construct small mounds of earth and debris on which to defecate.

URINE AND OTHER SCENT-MARKING BEHAVIORS: Muskrats have paired musk glands near the anus that are capable of releasing a conspicuous odor (Willner 1980). During the breeding season, scent is deposited around latrines, trails, and den sites (Errington 1963).

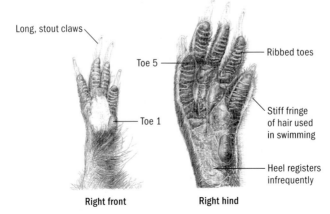

Common Muskrat feet.

Loping Muskrat.   Walking Muskrat.

The front and hind tracks of a Muskrat along the Sacramento River. Note the claw from toe 1 in the front right track (below) and the ribbed digital pads in both the front and hind (above) tracks.

From top to bottom, hind, front, hind tracks of a Muskrat in soft mud.

A Muskrat latrine.

WOODRATS *Neotoma* spp.

TRACKS: Front: ³/₈–1 in. long (1.1–2.5 cm) × ⁷/₁₆–¹⁵/₁₆ in. wide (1.1–2.4 cm)

Tiny. Plantigrade. Asymmetrical. Five toes in classic rodent structure: toe 1 is greatly reduced but registers clearly in good substrate. Toes 2 and 5 angle toward the sides, and toes 3 and 4 point forward. The digital pads are distinctly bulbous and register separately from the palm. Some of the metacarpal pads are fused, three distinct palm pads are present; two additional pads, the metacarpal and carpal, form the heel at the posterior edge of the track, and often show. Claws and heel may or may not register. The front tracks are round, and only slightly smaller than the hind tracks. Woodrat tracks are very similar to deer mice tracks, but larger. Note: Refer to Roof Rat and Brown Rat accounts for a comparison with those species.

Hind: ⁷/₈–1¼ in. long (1.6–3.2 cm) × ⁷/₁₆–1 in. wide (1.1–2.5 cm)

Tiny. Plantigrade. Slightly asymmetrical. Five toes in a classic rodent structure: toes 1 and 5 angle to the sides, and toes 2, 3, and 4 point forward. The digital pads are distinctly bulbous. Some metatarsal pads are fused to form larger pads, but four palm pads should be distinct, with two additional heel pads at the posterior of the track that often register when woodrats are bounding in good substrates. The posterior heel pads are distinctly asymmetrical and may form the three points of a triangle. Claws and heel may or may not register.

TRAIL: Bound: Strides: 4–8 in. (10.2–20.3 cm)
Trail widths: 2¹/₈–3¹/₈ in. (5.4–7.9 cm)

Walk: Strides: 1⁵/₈–3½ in. (4.1–8.9 cm)
Trail widths: 1⁵/₈–2¼ in. (4.1–5.7 cm)

NOTES:

• There are five species of woodrats in California: the larger species are the Bushy-tailed (*Neotoma cinerea*), Big-eared (*Neotoma macrotis*), and Dusky-footed (*Neotoma fuscipes*) woodrats. The smaller are the White-throated (*Neotoma albigula*) and the Desert (*Neotoma lepida*) woodrats. The Desert Woodrat is half the size and weight of the larger Big-eared Woodrat, with which it shares habitat in southern California.

• Woodrats bound across exposed areas or when threatened, but tend to walk or trot in or near cover when foraging, scent-marking, or traveling.

• Woodrats form substantial runs, leading from nests or burrows to regular harvesting areas.

• Woodrats are competent climbers and often live and harvest above ground.

SCAT: ¹/₈–³/₁₆ in. (0.3–0.5 cm) diameter, ¼–⁵/₈ in. (0.6–1.6 cm) long

Where there are woodrats, you will find ample amounts of scats and other signs; great accumulations will be found along travel routes, and in and around burrows and nests. Scat latrines can be impressive, with small, tubular scats flowing forth from beneath rocks like rivers, covering several square feet and accumulating several inches deep. Woodrats also defecate within their nests and then push them out when accumulations become an obstacle. Scats are tubular pellets, much longer than wide, and are similar across species. Woodrats also excrete a softer, dark, tarlike scat, which they often post along travel routes and near nests. These scent posts are used time and again, and the black goo will accumulate impressively over time.

Vestal (1938) reported that it is possible to differentiate scats of adult male and female Dusky-footed Woodrats by scat size. Female scats averaged 0.36 in. (9.1 mm) long by 0.14 in. (3.6 mm) diameter, while male scats averaged 0.47 in. (11.9 mm) long by 0.19 in. (4.8 mm) diameter. One captive female averaged 144 droppings per 24-hour period (six per hour) over a 168-hour period. In a population of captive woodrats, urine was deposited away from the nest.

URINE AND OTHER SCENT-MARKING BEHAVIORS: Woodrats also use urine for scenting purposes, and in dry climates the spots where woodrats urinate most will become white with calcareous deposits. White streaks also form below or adjacent to nests in rock cliffs, and can be confused with the white stains created by roosting raptors; however, they are often horizontal in orientation rather than vertical.

During the breeding season male abdominal ventral glands swell and probably communicate much about sexual status, dominance, and territoriality. Males rub their ventral glands on rocks, sticks, nest entrances, and travel routes, sometimes exhibiting a perineal drag, where they splay their hind legs and literally drag their hind ends to deposit scent marks (Howe 1977). Males also mark as an act of dominance, marking more frequently near the nests of subordinates than their own. Subordinate males mark less frequently than dominants, and avoid marking near the nests of dominants.

Females, more often than males, roll to scent-mark, digging up earth with their forelimbs and then rubbing their cheek, shoulder, hind leg, and sides into the ground. Males scent with their ventral glands in response to finding the ventral rubs of other males, but roll to mark when they encounter the rolls of females.

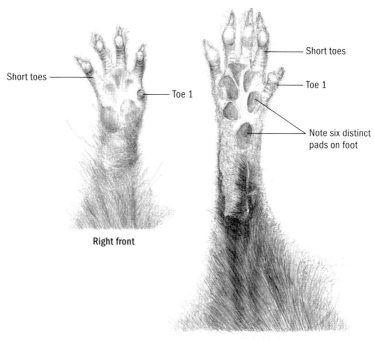

Short toes

Short toes

Toe 1

Toe 1

Note six distinct
pads on foot

Right front

Right hind

Big-eared Woodrat feet.

Bounding woodrat.   Walking woodrat.

The front and hind tracks of a Desert Woodrat.

The walking trails of a Western Toad (*left*) and Big-eared Woodrat in the Los Padres National Forest.

The mysterious black, tarlike secretions of an unknown woodrat species along a regular travel route.

The latrine of a Big-eared Woodrat in the crotch of an oak tree near Monterey.

The accumulated calcareous deposits of a Bushy-tailed Woodrat near a nest.

DEER MOUSE (WHITE-FOOTED MOUSE)   *Peromyscus* spp.

Plate 1
Plate 13
Plate 87

TRACKS: Front: ¼–½ in. long (0.6–1.3 cm) × ⁵⁄₁₆–⁹⁄₁₆ in. wide (0.8–1.4 cm).
Track and trail parameters for the California (old name Parasitic)
Deer Mouse (*P. californicus*) may be larger.

Tiny. Plantigrade. Asymmetrical. Five toes in classic rodent structure:
toe 1 is greatly reduced but visibly registers in good substrate. Toes
2 and 5 angle toward the sides, and toes 3 and 4 point forward.

The digital pads are bulbous and typically register separately from the palm pads. Some of the metacarpal pads are fused, but there are three distinct palm pads; an additional metacarpal pad and the carpal pad form the heel at the posterior edge of the track, and often show. Claws and heel may or may not register. The front tracks are round, and only slightly smaller than the hind tracks.

Hind: ¼–⅝ in. long (0.6–1.6 cm) × ⁵/₁₆–⁹/₁₆ in. wide (0.8–1.4 cm). Track and trail parameters for the California Deer Mouse (*P. californicus*) may be larger.

Tiny. Plantigrade. Asymmetrical. Five toes in a classic rodent structure: toes 1 and 5 angle to the sides, and toes 2, 3, and 4 point forwards. The digital pads are bulbous. Some metatarsal pads are fused to form larger pads, but four palm pads should be distinct, with two additional heel pads at the posterior of the track; the impressions of the central palm pads often form an "¬" and "·", the dot forming the corner of the rectangle if the "¬" and "·" were combined. Note that the rectangle created by the "¬"–"·" combination is wider than long, whereas in voles it is longer than wide. Claws and heel may or may not register.

TRAIL: Walk: Strides: 1⅜–2⅛ in. (3.5–5.4 cm)
                       Trail widths: 1⅛–1¾ in. (2.9–4.4 cm)

          Bound: Strides: 4–20 in. (10.2–50.8 cm)
                       Trail widths: 1⅜–1¾ in. (3.5–4.4 cm)
                       Group lengths: 1⅛–2⅛ in. (2.9–5.4 cm)

NOTES:

- There are six species of deer mice in California, and in general, their sign is difficult to distinguish from each other. The California Deer Mouse can be up to twice as large as all the other species.

- Mice bound when traveling, walking only when foraging or under thick, protective cover.

- They are capable swimmers and competent climbers; mice nest and forage at every height in vegetative strata.

- In deep snow, deer mice continue to move on top of the snow layer more than other small mammals. In softer conditions, the front and hind tracks blend together to form a 2 × 2 bound. Tail drag may or may not be obvious.

SCAT: ¹/₃₂–¹/₁₆ in. (0.1–0.2 cm) diameter, ³/₃₂–⁷/₃₂ in. (0.2–0.6 cm) long

          Scats are irregular, wrinkled, twisted, and tapered. Scats seem to be dropped at random and can be found wherever mice have spent time, accumulating along travel routes and where actively foraging. However, mice do not defecate in their nests, instead scattering them about their home ranges and forming latrines near the nest (Eisenberg 1962). In captivity they tend to defecate in a certain corner of their cages, which is where they also retreat to groom themselves.

URINE AND OTHER SCENT-MARKING BEHAVIORS: Mice urinate wherever they wander, "painting" their respective ranges with their own scent. Extensive work with laboratory mice (*Mus musculus*) shows that dominant mice urinate constantly as they wander. Subordinates mark less frequently and in fewer areas.

The California Deer Mouse, which is more territorial than other *Peromyscus*, has especially large preputial glands (located in the anal region), and scent-marks by dragging its backside along the ground and on other objects like sticks and rocks (Eisenberg 1962). Anal-dragging is also exhibited by deer mice and is common when the mouse is exposed to a new environment and immediately after a fight.

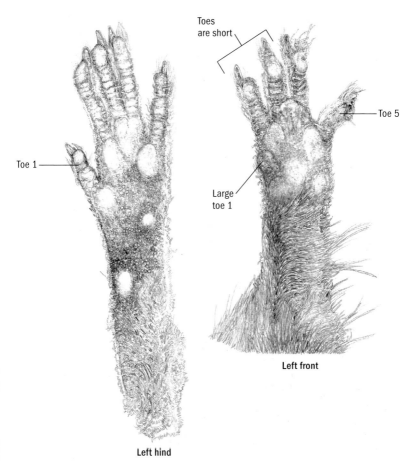

Toes
are short

Toe 5

Toe 1

Large
toe 1

Left front

Left hind

North American Deer Mouse feet.

Numerous trails of deer mice bounding back and forth between two holes in the snow layer.

Bounding deer mouse.

Deer mice (also collectively called white-footed mice) tracks in a bound.

WESTERN HARVEST MOUSE   *Reithrodontomys megalotis*

TRACKS: Front: ¼–³/₈ in. long (0.6–1 cm) × ¼–⁵/₁₆ in. wide (0.6–0.8 cm)

Tiny. Plantigrade. Asymmetrical, often much more so than in other mice species. Five toes in classic rodent structure: toe 1 is greatly reduced. Toes 2 and 5 point toward the sides, and toes 3 and 4 point forward. The digital pads are bulbous, though less so than the pads of White-footed Mice. Some of the metacarpal pads are fused, but there are three distinct palm pads in tracks; an additional metacarpal pad and the carpal pad form the heel at the posterior edge of the track, which may or may not show. Claws may or may not register. The front tracks are of similar dimensions to the hind tracks.

Hind: ¼–½ in. long (0.6–1.3 cm) × ⁷/₃₂–⁵/₁₆ in. wide (0.6 –0.8 cm)

Tiny. Plantigrade. Asymmetrical. Five long toes in a classic rodent structure: toes 1 and 5 point to the sides, and toes 2, 3, and 4 all point forward together. However, toe 1 is slightly longer and situated further to the posterior in the track than in other mouse species. Toe 5 occasionally points forward and, though shorter, may sit close to toe 4. Some metatarsal pads are fused to form larger pads, but four palm pads should be distinct, with two additional heel pads at the posterior of the track. Claws and heel may or may not register. The position and shape of toe 1 in hind tracks help differentiate harvest mice from other mice species.

TRAIL: Bound: Strides: 2–9 in. (5.1–22.9 cm)
             Trail widths: 1–1³/₈ in. (2.5–3.5 cm)
             Group lengths: 1–1¾ in. (2.5–4.4 cm)

NOTES:

- Move about in a bound, walking only when foraging or under thick, protective cover.
- Competent climbers; mice may be found at any height in vegetative strata.

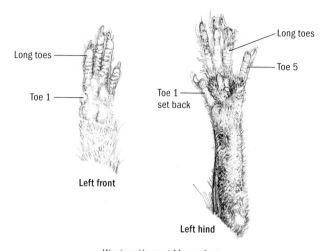

Western Harvest Mouse feet.

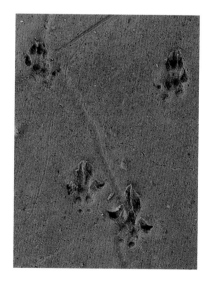

The bounding trail of a Western Harvest Mouse.

HISPID COTTON RAT

Sigmodon hispidus

TRACKS: Front: $7/16$–$5/8$ in. long (1.1–1.6 cm) × $7/16$–$5/8$ in. wide (1.1–1.6 cm)

Tiny. Plantigrade. Asymmetrical. Five toes in classic rodent structure: toe 1 is greatly reduced and does not typically register. Toes 2 and 5 angle toward the sides, and toes 3 and 4 point forward. Digital pads typically connect to the metacarpal pads, or come close. Some of the metacarpal pads are fused, but there are three distinct palm pads; an additional metacarpal pad and the carpal pad form the heel at the posterior edge of the track, and they may or may not register. Claws too may or may not register. The front tracks are smaller than the hind ones.

Hind: $9/16$–$7/8$ in. long (1.4–2.2 cm) × $9/16$–$11/16$ in. wide (1.4–1.7 cm)

Tiny. Plantigrade. Slightly asymmetrical. Five long, slender toes in a classic rodent structure: toes 1 and 5 angle to the sides, and toes 2, 3, and 4 point forward. Metatarsal pads are fused to form larger pads, but four lobed pads should be distinct, with two additional heel pads at the posterior of the track. Claws are small and may or may not register. Heel may or may not register.

TRAIL: Walk: Strides: $1^5/8$–$2^1/4$ in. (4.1–5.7 cm)
Trail widths: $1^3/8$–$1^7/8$ in. (3.5–4.8 cm)

Trot: Strides: 4–5 in. (10.2–12.7 cm)
Trail widths: $1^1/8$–$1^5/8$ in. (2.9–4.1 cm)

NOTES:

- Tracks, though larger, may look similar to those of voles.
- Thin, straighter toes help differentiate this species from other California rats.
- Occur in the extreme southeastern corner of California.
- They walk when foraging or exploring, and trot or bound across open areas and when feeling threatened or exposed.

SCAT: $1/32$–$1/8$ in. (0.1–0.3) diameter, $5/32$–$5/16$ in. (0.4–0.8 cm) long.

Gregory and Cameron (1989) discovered that Hispid Cotton Rats selected sites to deposit scat and urine according to their social rank within their local population, as well as varied in their reaction to the scats and urine of other rats in their range depending upon their social rank.

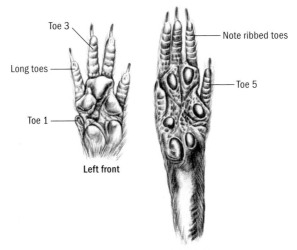

Toe 3

Long toes

Toe 1

Left front

Note ribbed toes

Toe 5

Left hind

Hispid Cotton Rat feet.

A bounding Hispid Cotton Rat. Photo by Jonah Evans.

Walking Hispid Cotton Rat.

The trotting trail of a Hispid Cotton Rat.

Exotic Rats and Mice: Family Muridae

Rats

BROWN RAT (NORWAY RAT) *Rattus norvegicus*

TRACKS: Front: $^{1}/_{2}$–$^{7}/_{8}$ in. long (1.3–2.2 cm) × $^{1}/_{2}$–$^{3}/_{4}$ in. wide (1.3–1.9 cm)

ROOF RAT (BLACK RAT) *Rattus rattus*

TRACKS: Front: $^{3}/_{8}$–$^{11}/_{16}$ in. long (1–1.7 cm) × $^{1}/_{2}$–$^{11}/_{16}$ in. wide (1.3–1.7 cm)

Tiny to small. Plantigrade. Asymmetrical. Five toes in classic rodent structure: toe 1 is greatly reduced. Toes 2 and 5 point almost directly toward the sides, and toes 3 and 4 point forward. The toes are slightly shorter in Roof Rats than in Brown Rats. The digital pads generally connect to the metacarpal region in both species but may register separately more often in the Roof Rat. Some of the metacarpal pads are fused, but there are three distinct palm pads; an additional metacarpal pad and the carpal pad form the heel at the posterior edge of the track, and they may or may not show. Both the digital and metacarpal pads are significantly smaller in the Brown Rat than the Roof Rat, which exhibits bulbous pads more akin to those in woodrat tracks. Claws tend not to register. The front tracks are smaller than the hind ones. The front track is also more elongated than round, as opposed to the front tracks of woodrats.

Brown Rat (Norway Rat)

Hind: $^{3}/_{4}$—$1^{3}/_{8}$ in. long (1.9–3.5 cm) × $^{5}/_{8}$–$1^{3}/_{16}$ in. wide (1.6–3 cm)

Roof Rat (Black Rat)

Hind: $^{1}/_{2}$–1 in. long (1.9–2.5 cm) × $^{5}/_{8}$–1 in. wide (1.6–2.5 cm)

Small. Plantigrade. Slightly asymmetrical. Five long toes in a classic rodent structure: toes 1 and 5 point straight to either side, and toes 2, 3, and 4 point forward. The toes of Roof Rats are proportionately shorter than those of Brown Rats, especially apparent in a comparison of toes 2–4. The metatarsal pads are fused to form larger pads, but four lobed pads should be distinct, as well as two additional heel pads at the posterior of the track if the entire foot came into contact with the ground. Both the digital and metatarsal pads are significantly smaller in the Brown Rat than the Roof Rat, which exhibits bulbous pads more akin to those in woodrat tracks. Claws may or may not register. The toes are long and slender, and the greater space between the tips of the toes and the palm helps differentiate exotic rat tracks from those of native woodrats.

TRAIL: Walk: Strides: 3–5 in. (7.6–12.7 cm)
Trail widths: $1^{5}/_{8}$–$2^{3}/_{4}$ in. (4.1–7 cm)

Trot: Strides: 4–6 in. (10.2–15.2 cm)
            Trail widths: 2½–2⁵/₈ in. (6.4–6.7 cm)

Bound: Strides: 6–21 in. (15.2–53.3 cm)
            Trail widths: 1⁵/₈–3¼ in. (4.1–8.3 cm)
            Group lengths: 2³/₈–4 in. (6–10.2 cm)

2 × 2:   Strides: 7¼–10 in. (18.4–25.4 cm)
            Trail widths: 1⁵/₈–2 in. (4.1–5.1 cm)
            Group lengths: 1³/₈–2¼ in. (3.5–5.7 cm)

NOTES:

• Both Brown and Roof rats are exotic species accidentally introduced to California and now widespread. The Brown Rat is often twice the size of the Roof Rat, and is less abundant. Roof Rats are incredibly widespread along coastal and central California.

• According to mammalogist Paul Collins, curator for the Santa Barbara Museum of Natural History and trapper of thousands of rats in southern California, the habitats of Roof and Brown rats are distinctive and aid in their identification. Where vegetation is the predominant habitat, or where vegetation is mixed with concrete, Roof Rats predominate. Where habitats are completely concrete, Brown Rats dominate. In Santa Barbara, for instance, Roof Rats are too numerous to count, and Brown Rats are rare and difficult to find.

• In terms of toe length (in proportion to track size), and the robustness of the digital and metacarpal/metatarsal pads, woodrats and Brown Rats are opposite ends of the spectrum. Woodrats have the shortest toes and Brown Rats the longest. Woodrats have round, bulbous digital and palm pads, and Brown Rats have slender, small pads. The Roof Rat, however, lies exactly in the middle of these two species when discussing these track features, and so their sign is much more easily confused with that of woodrats.

• Rats walk when foraging and exploring, and trot to travel under cover. Rats bound when traveling in exposed areas or when threatened.

• Rats are both competent swimmers and climbers.

• In deep snow, rats tend to use either a 2 × 2 lope or 2 × 2 bound. Tail drag is often obvious.

SCAT: ¹/₈–¼ in. (0.3–0.6 cm) diameter, ¼–¹⁵/₁₆ in. (0.6–2.4 cm) long

Scats accumulate wherever rats spend time, especially near nest sites, along travel routes, and wherever they have been foraging.

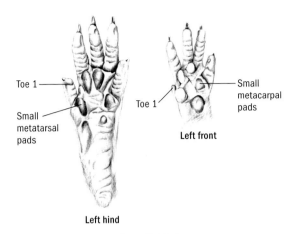

Toe 1

Small
metacarpal
pads

Toe 1

Small
metatarsal
pads

Left front

Left hind

Brown Rat feet.

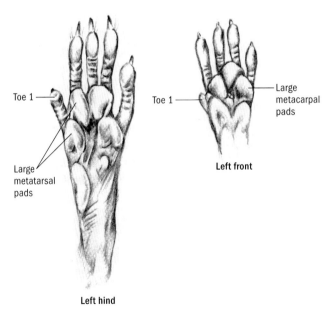

Toe 1

Large
metacarpal
pads

Toe 1

Large
metatarsal
pads

Left front

Left hind

Roof Rat feet.

Bounding trail of a Brown Rat.

Walking trail of a Brown Rat.

The bounding trail of a Brown Rat.

Roof Rat (Black Rat) bounding in shallow mud in the northern Sacramento Valley, where they are numerous.

Roof Rat in an understep trot. There is the hind track of a Western Gray Squirrel in the corner of the photograph.

Porcupine: Family Erethizontidae

NORTH AMERICAN PORCUPINE   *Erethizon dorsatum*

TRACKS: Front: 2¼–3⅜ in. long (5.7–8.6 cm) × 1¼–1⅞ in. wide (3.2–4.8 cm)

Medium. Plantigrade. Asymmetrical. Four very long toes. Toe 1 is absent altogether, and the remaining toes register together, and curve inward. The digital pads of porcupines only register in deep substrate; typically they do not register at all, and it is only the tips of the massive claws that are visible in shallow substrates. The metacarpal pads are fused together to form a unique and somewhat misshapen palm, with a characteristic pebbly surface, like the surface on a basketball. The claws are large and a prominent feature in tracks, registering so far beyond the palm that they might be overlooked. The front tracks are significantly smaller than the hind ones.

Hind: 2¾–4 in. long (7–10.2 cm) × 1¼–2 in. wide (3.2–5.1 cm); young are 2½ in. long (6.4 cm)

Medium. Plantigrade. Slightly asymmetrical. Five long toes. Toe 1 is much smaller than toes 2–5 and may not register well. The digital pads are typically absent, unless the tracks are in deep substrate; instead, look for the claw marks well in front of the palm. The palm and heel pads are fused together into a single oval palm. The exposed skin is uniquely pebbly, like the surface of a basketball. The claws and heel are both prominent in tracks.

TRAIL: Walk: Strides: 6–10½ in. (15.2–26.7 cm); young 4–5 in. (10.2–12.7 cm)

Trail widths: 5–9 in. (12.7–22.9 cm); young 3–4 in. (7.6–10.2 cm)

NOTES:

- Walk everywhere, but are occasionally startled into a short-lived loping gait.
- Competent climbers, and often found feeding or resting in trees, far above the ground.
- Direct-register walk in snow, and plow through deep snow, creating a deep furrow.
- Make regular runs between their dens and feeding areas, apparent both in snow and leaf litter.

SCAT: ¼–½ in. (0.6–1.3 cm) diameter, ½–1¼ in. (1.3–3.2 cm) long

Porcupine scat can be found wherever there are porcupines, accumulating where they feed, dropped wherever they walk, and especially thick in crevices and hollows (Cahalane 1961), or even basements, where they rest. In fact, scat accumulations may be so high at resting places, porcupines may have to burrow through their own excrement to exit and enter; look for rivers of scat flowing from rock ledges or tree hollows where they hole up in winter. Some suggest this behavior may aid in providing shelters with insulation; winter scat is composed completely of tiny wood chips.

Scats are irregular small tubes, and "pellets." Rounded or pointy ends, and most scats curve over their length—an asymmetry that helps differentiate them from deer scats. Scats may also be linked and form chains, or occasionally have a groove running down one side.

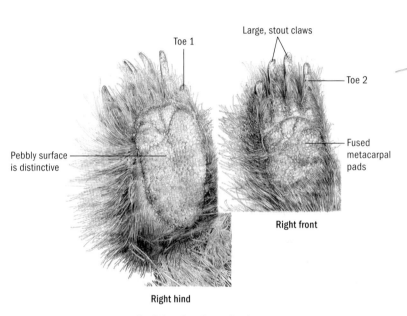

North American Porcupine feet.

Somewhat unusual North American Porcupine tracks, in that the digital pads for both the left hind foot (*left*) and left front foot are visible. More often you only see the claws.

Walking trail of a North American Porcupine.

The walking trail of a North American Porcupine in fine sand, where the dragging quills have also left sign. The tracks of a kangaroo rat can be found in the lower right corner.

The overflowing latrine or den of a North American Porcupine inhabiting rock ledges.

Nutria: Family Myocastoridae

COYPU (NUTRIA)   *Myocastor coypus*

Plate 31
Plate 97

TRACKS: Front: 1½–2⁷⁄₈ in. long (3.8–7.3 cm) × 1³⁄₈–2½ in. wide (3.5–6.4 cm)

Medium. Plantigrade. Asymmetrical. Five toes in classic rodent structure: toe 1 is reduced but clawed, and registers well in tracks. Toes 2 and 5 angle toward the sides, and toes 3 and 4 point forward. Some of the metacarpal pads are fused; an additional metacarpal pad and the carpal pad form the heel at the posterior edge of the track, and they may or may not show. Claws long, sharp, and prominent in tracks. The front tracks are significantly smaller than the hind tracks. The front tracks of Nutrias resemble the front tracks of Muskrats but are larger.

Hind: 2½–5⁷⁄₈ in. long (6.4–14.9 cm) × 1⁷⁄₈–3⁷⁄₈ in. wide (4.8–9.8 cm)

Medium to large. Plantigrade. Asymmetrical. Five long toes. Toes 1 and 2 are not always obvious in tracks. Distal webbing may register between all toes, except toes 4 and 5. The metatarsal pads are partially fused, yet distinct in tracks. The claws are large, short, and considerably sharper than those of beavers. The heels of the hind tracks rarely register when Nutria are walking, and so if you see a deep, rounded heel in a trail, you've more likely found a beaver track. Young Nutria trails can easily be confused with those of Muskrats.

TRAIL: Walk/overstep walk: Strides: 5–13 in. (12.7–33 cm)
Trail widths: 3¾–9¼ in. (9.5–23.5 cm)

Bounds: Strides: 10–30 in. (25.4–76.2 cm)
Trail widths: 7–10½ in. (17.8–26.7 cm)
Group lengths: 6–15 in. (15.2–38.1 cm)

NOTES:

- Explore areas in a walk, but quickly switch to a bound when threatened or alarmed. Large animals tend toward lopes rather than bounds.

- Wetland species, originally from South America, that were released in many areas of the United States. There have been attempts to eradicate them in California, and their signs are becoming increasingly rare in the state.

- Differentiating young Nutrias from Muskrats can be challenging. Examine the hind tracks; the presence of webbing will confirm Nutria, while a "shelf" of stiff hairs surrounding the track will confirm Muskrat.

SCAT: ³/₁₆–½ in. (0.5–1.3 cm) diameter, ¾–1½ in. (1.9–3.8 cm) long

Nutria scat is tubular and very symmetrical. The clear parallel grooves that run their lengths make these scats easy to identify in the field. Scats are sprinkled wherever Nutria move and are often seen floating in water sources they use and inhabit. Grass clumps and elevated surfaces are also used repeatedly, and accumulations will be found at such locations. Like rabbits and beavers, Nutrias engage in coprophagy, ingesting their scats to better absorb plant nutrients (Gosling 1979).

Walking trail of a Nutria (Coypu).

Bounding trail of a Nutria (Coypu).

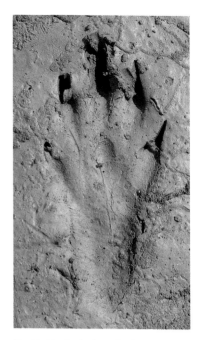

The characteristic walking trail of an adult Nutria (Coypu).

The hind track of a bounding Nutria. Note the complete registration of the heel, which does not occur when Nutrias are walking.

BIBLIOGRAPHY

Anderson, D. 2001. The need to get the basics right in wildlife field studies. *Wildlife Society Bulletin* 29:1294–1297.

Armitage, K. B. 1976. Scent marking by yellow-bellied marmots. *Journal of Mammalogy* 57:583–584.

———. 2003. Marmots. In *Wild mammals of North America: biology, management, and conservation*. 2d ed., ed. G. Feldhamer, B. Thompson, and J. Chapman, 188–211. Baltimore: Johns Hopkins University Press.

Arnold, J. F., and H. G. Reynolds. 1943. Dropping of Arizona and antelope jack rabbits and the "pellet census." *Journal of Wildlife Management* 7:322–327.

Asa, C. S., E. K. Peterson, U. S. Seal, and L. D. Mech. 1985. Deposition of anal-sac secretions by captive wolves (*Canis lupus*). *Journal of Mammalogy* 66:89–93.

Bailey, T. N. 1974. Social organization in a bobcat population. *Journal of Wildlife Management* 38:435–446.

Bang, P., and P. Dahlstrom. 1972. *Collins guide to animal tracks and signs*, translated and adapted by G. Vevers, 1974. London: Collins Presss.

Barja, I., and R. List. 2006. Faecal marking behaviour in ringtails (*Bassariscus astutus*) during the non-breeding period: spatial characteristics of latrines and single faeces. *Chemoecology* 16:219–222.

Ben-David, M., R. T. Bowyer, L. K. Duffy, D. D. Roby, and D. M. Schell. 1998. Social behavior and ecosystem processes: river otter latrines and nutrient dynamics of terrestrial vegetation. *Ecology* 79:2567–2571.

Ben-David, M., G. M. Blundell, J. W. Kern, J. A. K. Maier, E. D. Brown, and S. C. Jewett. 2005. Communication in river otters: creation of variable response sheds for terrestrial communities. *Ecology* 86:1331–1345.

Bowyer, R. T., and D. W. Kitchen. 1987. Significance of scent-marking by Roosevelt elk. *Journal of Mammalogy* 68:418–423.

Brinck, C., R. Gerell, and G. Odham. 1978. Anal pouch secretion in mink *Mustela vison*. Chemical communication in Mustelidae. *Oikos* 30:68–75.

Burst, T. L., and M. R. Pelton. 1983. Black bear mark trees in the Smoky Mountains. *Bears: Their Biology and Management* 5:45–53.

Byers, J. 2003. Pronghorn. In *Wild mammals of North America: biology, management, and conservation*. 2d ed., ed. G. Feldhamer, B. Thompson, and J. Chapman, 998–1009. Baltimore: Johns Hopkins University Press.

Cahalane, V. H. 1961. *Mammals of North America*. New York: Macmillan.

Carraway, L. N., and B. J. Verts. 1993. *Aplodontia rufa.* The American Society of Mammalogists: Mammalian Species No. 431:1–10.

Clark, J., T. Hon, K. D. Ware, and J. H. Jenkins. 1987. Methods for evaluating abundance and distribution of river otters in Georgia. *Proceedings from the Annual Conference of the Southeastern Association of Fish and Wildlife Agencies* 41:358–364.

Coppedge, B. R., and J. H. Shaw 1997. Effects of horning and rubbing behavior by bison (*Bison bison*) on woody vegetation in a tallgrass prairie landscape. *American Midland Naturalist* 138:189–196.

Darden, S. K., L. K. Steffensen, and T. Dabelsteen. 2008. Information transfer among widely spaced individuals: latrines as a basis for communication networks in the swift fox? *Animal Behaviour* 75:425–432.

De Angelo, C., A. Paviolo, and M. S. Di Bitetti. 2010. Traditional versus multivariate methods for identifying jaguar, puma, and large canid tracks. *Journal of Wildlife Management* 74:1141–1153.

DeBruyn, T. 1999. *Walking with bears.* New York: The Lyons Press.

Devenport, L., J. Devenport, and C. Kokesh. 1999. The role of urine marking in the foraging behaviour of least chipmunks. *Animal Behaviour* 57:557–563.

Eisenberg, J. F. 1962. Studies on the behavior of *Peromyscus maniculatus gambelli* and *Peromyscus californicus parasiticus.* *Behaviour* 19:177–207.

———. 1963. The behavior of heteromyid rodents. *University of California Publications on Zoology* 69:1–100.

Elbroch, M. 2003. *Mammal tracks and sign: a guide to North American species.* Mechanicsburg, Pennsylvania: Stackpole Books.

Elbroch, M., and E. Marks. 2001. *Bird tracks and signs: a guide to North American species.* Mechanicsburg, Pennsylvania: Stackpole Books.

Elbroch, M., T. H. Mwampamba, M. Zylberberg, M. J. Santos, L. Liebenberg, J. Minye, C. Mosser, and E. Reddy. In press. The value of local experts to conservation research. *Conservation Biology.*

Errington, P. L. 1963. *Muskrat populations.* Ames: Iowa State University Press.

Evans, J. W., C. A. Evans, J. M. Packard, G. Calkins, and M. Elbroch. 2009. Determining observer reliability in counts of river otter tracks. *Journal of Wildlife Management* 73:426–432.

Fadem, B. H., and E. A. Cole. 1985. Scent-marking in the grey short-tailed opossum (*Monodelphis domestica*). *Animal Behaviour* 33:730–738.

Fair, J., with L. Rogers. 1990. *The great American bear.* Minocqua, Wisconsin: Northword Press.

Ferkin, M. H., S. G. Mech, and G. Paz-Y-Miño C. 2001. Scent marking in meadow voles and prairie voles: a test of three hypotheses. *Behaviour* 138:1319–1336.

Ferron, J. 1977. Le comportement de marquage chez le spermophile a mante dorae (*Spermophilus lateralis*). *Nat. Can. (Quebec),* 104:407–418.

———. 1983. Scent marking by cheek rubbing in the northern flying squirrel (*Glaucomys sabrinus*). *Canadian Journal of Zoology* 61:2377–2380.

Ferron, J., and J. Ouellet. 1989. Behavioral context and possible function of scent marking by cheek rubbing in the red squirrel (*Tamiasciurus hudsonicus*). *Canadian Journal of Zoology* 67:1650–1653.

Full, R. J., and M. S. Tu. 1991. Mechanics of six-legged runners. *Journal of Experimental Biology* 148:129–146.

Galloway, A.W.E., M. T. Tudor, and W. M. Vander Haegen. 2006. The reliability of citizen science: a case study of Oregon white oak stand surveys. *Wildlife Society Bulletin* 34:1425–1429.

Gamberg, M., and J. Atkinson. 1988. Prey hair and bone recovery in ermine scats. *Journal of Wildlife Management* 52:657–660.

García, K. P., J. C. Ortiz, M. Vidal, and J. R. Rau. 2010. Morphometrics of the tracks of *Puma concolor:* is it possible to differentiate the sexes using measurements from captive animals? *Zoological Studies* 49:577–582.

Gerell, R. 1968. Food habits of the mink (*Mustela vison*) in Sweden. *Viltrevy* 5:199–211.

Gese, E., and R. Ruff. 1997. Scent-marking by coyotes, *Canis latrans*: the influence of social and ecological factors. *Animal Behaviour* 54:1155–1166.

Gorman, M. L., and Stone, R. D. 1990. Mutual avoidance by European moles *Talpa europaea*. In *Chemical signals in vertebrates.* Vol. 5, ed. D.W. Macdonald, D. Müller-Schwartze, and S. E. Natynczuk, 367–377. Oxford: Oxford University Press.

Gosling, L. M. 1979. The twenty-four hour activity cycle of captive Coypus (*Myocastor coypus*). *Journal of Zoology* (London) 187:341–367

Green, G. I., and D. J. Mattson. 2003.Tree rubbing by Yellowstone grizzly bears *Ursus arctos*. *Wildlife Biology* 9:1–9.

Green, J. S., and J. T. Flinders. 1980. *Brachylagus idahoensis*. The American Society of Mammalogists: Mammalian Species No. 125:1–4.

Gregory, M. J., and G. N. Cameron. 1989. Scent communication and its association with dominance behavior in the Hispid Cotton Rat (*Sigmodon hispidus*). *Journal of Mammalogy* 70:10–17.

Grinnell, J., J. S. Dixon, and J. M. Linsdale. 1937. *Fur-bearing mammals of California: their natural history, systematic status, and relations to man.* Berkeley: University of California Press.

Halfpenny, J., M. Walla-Murphy, E. Rogers, and D. Tiller. 2010. *Track plates for mammals: a how to manual and aid to footprint identification.* Gardiner, Montana: A Naturalist's World.

Hall, E. R. and K. R. Kelson. 1959. *The mammals of North America.* Vol. 2, 547–1083. New York: Ronald Press + 79.

Harmsen, B. J., R. J. Foster, S. M. Gutierrez, S. Y. Martin, and C. P. Doncaster. 2010. Scrape-marking behavior of jaguars and pumas. *Journal of Mammalogy* 91:1225–1234.

Harrington, F. H. and C. A. Asa. 2003. Wolf communication. In *Wolves: behavior, ecology and conservation,* ed. L. D. Mech and L. Boitani, 66–103. Chicago: The University of Chicago Press.

Hartman, G. D., and T. L. Yates. 2003. Moles (Talpidae). In *Wild mammals of North America: biology, management, and conservation.* 2d ed., G. Feldhamer, B. Thompson, J. Chapman, 30–55. Baltimore: Johns Hopkins University Press.

Henry, J. D. 1977. The use of urine marking in the scavenging behaviour of the red fox (*Vulpes vulpes*). *Behaviour* 61:82–105.

————. 1993. *How to spot a fox.* Shelburne, Vermont: Chapters Publishing Co.

Herzog, C. J., R. W. Kays, J. C. Ray, M. E. Gompper, W. J. Zielinski, R. Higgens, and M. Tymeson. 2009. Using patterns in track-plate footprints to identify individual fishers. *The Journal of Wildlife Management* 71:955–963.

Hildebrand, M., and G. Goslow. 2001. *Analysis of vertebrate structure.* 5th ed. New York: John Wiley and Sons.

Howe, R. J. 1977. Scent-marking behavior in three species of woodrats (*Neotoma*) in captivity. *Journal of Mammalogy* 58:685–688.

Irschick, D. J., and B. C. Jayne. 1999. Comparative three dimensional kinematics of the hindlimb for high-speed bipedal and quadrupedal locomotion of lizards. *Journal of Experimental Biology* 202:1047–1065.

Jameson, E. W., and Peeters, H. J. 2004. *Mammals of California.* Berkeley: University of California Press.

Karanth, K. U., J. Nichols, J. Seidensticker, J. L. David Smith, C. McDougal, A. J. Johnsingh, R. S. Chundawat, and V. Thapar. 2003. Science deficiency in conservation practice: the monitoring of tiger populations in India. *Animal Conservation* 6:141–146.

Kilham, B., and E. Gray. 2003. *Among the bears: raising orphaned cubs in the wild.* New York: MacMillan.

Kitchen, D. W. 1974. Social behavior and ecology of the pronghorn. *Wildlife Monographs* 38:3–96.

Kivett, V. K., J. O. Murie, and A. L. Steiner. 1976. A comparative study of scent-gland location and related behaviour in some northwestern neoarctic ground squirrel species (Sciuridae): an evolutionary approach. *Canadian Journal of Zoology* 54:1294–1306.

Koehler, G. M., G. H. Maurice, and S. H. Howard. 1980. Wolverine marking behavior. *Canadian Field-Naturalist* 94:339–341.

Koprowski, J. L. 1993. Sex and species biases in scent-marking by fox squirrels and eastern grey squirrels. *Journal of Zoology* 230:319–356.

Larivière, S., and F. Messier. 1996. Aposematic behaviour in the striped skunk, *Mephitis mephitis. Ethology* 102:986–992.

Laws, J. M. 2007. *The Laws field guide to the Sierra Nevada.* Berkeley: Heyday Books and the California Academy of Sciences.

Leahy, C. 1982. *The birdwatcher's companion.* New York: Gramercy Books.

Leslie, C.W. 1995. *Nature drawing—a tool for learning.* Dubuque, Iowa: Kendall Hunt.

Liebenberg, L. 1990. *The art of tracking: the origin of science.* Cape Town, South Africa: David Philip.

Liebenberg, L., A. Louw, and M. Elbroch. 2010. *Practical tracking: a guide to following footprints and finding animals.* Mechanicsburg, Pennsylvania: Stackpole Books.

Lindzey, F. G. 2003. Badger. In *Wild mammals of North America: biology, management, and conservation.* 2d ed., ed. G. Feldhamer, B. Thompson, and J. Chapman, 683–691. Baltimore: Johns Hopkins University Press.

Long, R. A., P. MacKay, J. Ray, and W. Zielinski. 2008. *Noninvasive survey methods for carnivores.* Washington: Island Press.

Macdonald, D. W. 1985. The carnivores: order Carnivora. In *Social odours in mammals*. Vol 2, ed. R. E. Brown and D. W. Macdonald, 619–722. Oxford: Oxford University Press.

Maehr, D. S. 1997. The comparative ecology of bobcat, black bear and Florida panther in south Florida. *Bulletin of the Florida Museum of Natural History* 40:1–176.

McManus, J. J. 1970. Behavior of captive opossums, *Didelphis-marsupialis-virginiana*. *American Midland Naturalist* 84:144–169.

McMillan, B. R., M. R. Cottam, and D. W. Kaufman. 2000. Wallowing behavior of American bison (*Bos bison*) in tallgrass prairie: an examination of alternate explanations. *American Midland Naturalist* 144:159–167.

Mech, D., U. Seal, and G. Delgivdice. 1987. Use of urine in snow to indicate condition of wolves. *Journal of Wildlife Management* 51:10–13.

Melquist, W. E., and M. G. Hornocker. 1983. Ecology of river otters in west central Idaho. *Wildlife Monographs* 83:3–60.

Müller-Schwarze, D., and S. Heckman. 1980. The social role of scent marking in beaver (*Castor canadensis*). *Journal of Chemical Ecology* 6:81–95.

Murie, O. 1954. *A field guide to animal tracks*. New York: Houghton Mifflin Co.

Murie, O, and M. Elbroch. 2005. *A field guide to animal tracks*. 3d ed. New York: Houghton Mifflin Co.

Muybridge, E. 1957. *Animals in motion*. New York: Dover Publications.

Nerbonne, J. F., and K. C. Nelson. 2008. Volunteer macroinvertebrate monitoring: tensions among group goals, data quality, and outcomes. *Environmental Management* 42:470–479.

Nolte, D. L., G. Epple, D. L. Campbell, and J. R. Mason. 1993. Response of mountain beaver to conspecifics in their burrow systems. *Northwest Science* 67:251–255.

Noordhuis, R. 2002. The river otter (*Lontra canadensis*) in Clarke County (Georgia, USA): survey, food habits and environmental factors. *IUCN Otter Specialist Group Bulletin* 19:75–86.

Ough, W. D. 1982. Scent marking by captive raccoons. *Journal of Mammalogy* 63:318–319.

Pearce, J. 1937. A captive New York weasel. *Journal of Mammalogy* 18:483–488.

Peterson, R. O., and P. Ciucci. 2003. The wolf as carnivore. In *Wolves: behavior, ecology and conservation*, ed. L. D. Mech and L. Boitani, 104–130. Chicago: The University of Chicago Press.

Polderboer, E. B., L. W. Kuhn, and G. O. Hendrickson. 1941. Winter and spring habits of weasels in central Iowa. *The Journal of Wildlife Management* 5:115–119.

Powell, R. A. 1981. *Martes pennanti*. Mammalian Species No. 156:1-6.

———. 1993. *The fisher: life history, ecology, and behavior*. 2d ed. Minneapolis: University of Minnesota Press.

Quinn, J. H. 2008. The ecology of the American badger *Taxidea taxus* in California: Assessing conservation needs on multiple scales. Ph.D. Diss., University of California, Davis.

Ryden, H. 1989. *Lily pond: four years with a family of beavers*. New York: William Morrow.

Samuel, W. M., M. J. Pybus, and A. Alan Kocan, eds. 2001. *Parasitic diseases of wild mammals.* Ames: Iowa State University Press.

Schwartz, C. W., and E. R. Schwartz. 2002. *The wild mammals of Missouri,* 2d ed. Columbia: University of Missouri Press.

Seidensticker IV, J. C., M. G. Hornocker, W. V. Wiles, and J. P. Messick. 1973. Mountain lion social organization in the Idaho Primitive Area. *Wildlife Monographs* 35:3–60.

Seton, E. T. 1958. *Animal tracks and hunter signs.* New York: Doubleday.

Shaw, H. 1979. Mountain lion field guide, Special report No. 9. Arizona Game and Fish Department.

Slauson, K. M., R. L. Truex, and W. J. Zielinski. 2008. Determining the gender of American martens and fishers at track plate stations. *Northwest Science* 82:185–198.

Smith, A. T. and M. L. Weston. 1990. *Ochotona princeps.* The American Society of Mammalogists: Mammalian Species No. 352:1–8.

Snyder, G. 1995. *A place in space: ethics, aesthetics, and watersheds.* Berkeley: Counterpoint.

Stains, H. J. 1956. The raccoon in Kansas. Miscellaneous Publication of the Museum of Natural History. University of Kansas, 10:1–76.

Stander, P., I. I. Ghau, D. Tsisaba, I. I. Oma, and I. I. Ui. 1997. Tracking and the interpretation of spoor: a scientifically sound method in ecology. *Journal of Zoology* 242:329–341.

Steiger, S., A. Fidler, and M. Valcu. 2008. Avian olfactory receptor gene repertoires: evidence for a well-developed sense of smell in birds? *Proceedings of the Royal Society of London, Biological Sciences* 1649:2309–2317.

Stewart, N., K. Scudder, and C. Southwick. 1982. Investigative response of pikas to odors produced by conspecifics. *Journal of Mammalogy* 63:690–694.

Svendsen, G. E. 1978. Castor and anal glands of the beaver (*Castor canadensis*). *Journal of Mammalogy* 59:618–620.

———. 1979. Territoriality and behavior in a population of pikas (*Ochotona princeps*). *Journal of Mammalogy* 60:324–330.

———. 1980. Patterns of scent-mounding in a population of beaver (*Castor-canadensis*). *Journal of Chemical Ecology* 6:133–148.

Sweeney, J. R., J. M. Sweeney, and S. W. Sweeney. 2003. Feral hog. In *Wild mammals of North America: biology, management, and conservation.* 2d ed., ed. G. Feldhamer, B. Thompson, and J. Chapman, 1164–1179. Baltimore: Johns Hopkins University Press.

Tarasoff, F. J. 1972. Comparative aspects of the hindlimbs of the river otter, sea otter, and harp seal. In *Functional anatomy of marine mammals*, ed. R. J. Harrison, 333–359. New York: Academic Press.

Trapp, G. R. 1978. Comparative behavioral ecology of ringtail and gray fox in southwestern Utah. *Carnivore* 1:3–32.

Vestal, E. H. 1938. Biotic relations of the wood rat (*Neotoma fuscipes*) in the Berkeley Hills. *Journal of Mammalogy* 19:1–36.

Weaver, J. 1993. Refining the equation for interpreting prey occurrence in gray wolf scats. *Journal of Wildlife Management* 57:534–538.

Weiner, J. 1995. *The beak of the finch: a story of evolution in our time.* New York: Vintage Books.

Wigley, T., and M. Johnson. 1981. Disappearance rates for deer pellets in the Southeast. *Journal of Wildlife Management* 45:251–253.

Williams, E. S., and I. K. Barker. 2000. *Infectious diseases of wild mammals,* 3d ed. Hoboken, New Jersey: Wiley-Blackwell Publishing.

Willner, G. R., G. A. Feldhamer, E. E. Zucker, and J. A. Chapman. 1980. *Ondatra zibethicus.* The American Society of Mammalogists: Mammalian Species No. 141:1–8.

Wilson, D. E. and D. M. Reeder, eds. 2005. *Mammal species of the world. A taxonomic and geographic reference.* 3rd ed. Baltimore: Johns Hopkins University Press.

Wilsson, L. 1971. Observations and experiments on the ethology of the European beaver (*Castor fiber* L.). *Viltrevy* 8:116–261.

Zielinski, W. and Kucera, T. 1996. *American marten, fisher, lynx and wolverine: survey methods for their detection.* Berkeley: Pacific Southwest Research Station, U. S. Forestry Service.

Zuercher, G. L., P. S. Gipson, and G. C. Stewart. 2003. Identification of carnivore feces by local peoples and molecular analyses. *Wildlife Society Bulletin* 31:961–970.

ART CREDITS

Photography by Mark Elbroch, unless otherwise noted

Mammal track and trail illustrations by Mark Elbroch

Mammal feet illustrations by Michael Kresky, excepting Virginia Opossum, American Badger, American Beaver, Hispid Cotton Rat, Brown Rat, and Roof Rat, which were reprinted from *The Wild Mammals of Missouri* by Schwartz and Schwartz (2002). Used with permission of the University of Missouri Press.

Bird tracks by Mark Elbroch

Bird, reptile, and amphibian feet by Michael Kresky

Insects and their trail illustrations by Michael Kresky

Reptile and amphibian trails by Mark Elbroch

INDEX

Page numbers in **bold** indicate the main discussion of the taxon. Page numbers followed by (fig.) indicate figures; plates are indicated by pl. followed by the plate number(s).

ABOUT THE AUTHORS

Mark Elbroch is author of numerous guides to wildlife tracking, including the award-winning *Mammal Tracks and Sign: A Guide to North American Species* and *Bird Tracks and Sign: A Guide to North American Species*. Mark was awarded a Senior Tracker Certificate by CyberTracker Conservation (www.cybertracker.org) in South Africa, the 17th given in more than 10 years of testing wildlife trackers, and the first awarded to a non-African. He carried the CyberTracker Tracker Evaluation system to North America, where it has been used to test observer reliability in wildlife projects and as an educational tool for varied institutions. Mark works as a biologist with a current focus on Pumas. His dissertation work at the University of California, Davis focused on the feeding ecology of Patagonia Pumas. He maintains the website http://wildlifetrackers.com.

Michael Kresky studied philosophy and art at Jacksonville University in Florida and then completed a teaching credential at UC Santa Barbara. He is a professional artist and professional natural historian, with specializations in wildlife tracking and native plants; he has taught natural history courses in venues as diverse as home school programs to university courses. Michael was tested and awarded a Track and Sign Specialist Certificate from CyberTracker Conservation in the woods he knows so well near his home in Santa Barbara (scoring higher than 100%!).

Jonah Evans currently lives in far west Texas working as a Non-game Biologist for Texas Parks and Wildlife conducting wildlife research on a variety of species. Jonah received an M.S. in Wildlife Science from Texas A&M University where he studied issues of observer reliability in a Texas Parks and Wildlife river otter track survey. He was tested in southern California and awarded a Track and Sign Specialist Certificate by CyberTracker Conservation, and then went on to learn and contribute to the system as an Evaluator. Jonah has worked on numerous wildlife projects, studying Lynx, river otter, Fisher, goshawk, elk, Black-tailed Prairie Dogs, and Golden-cheeked Warblers; he continues to work toward improving the tracking skills of biologists and the effectiveness of tracking-based research methods. Jonah maintains the website http://naturetracking.com.

Series Design:	Barbara Jellow
Design Enhancements:	Beth Hansen
Design Development:	Jane Tenenbaum
Indexer:	Nancy Newman
Composition:	P.M. Gordon Associates
Text:	9/10.5 Minion
Display:	ITC Franklin Gothic Book and Demi
Prepress:	Embassy Graphics
Printer and Binder:	Imago